The Chelation Answer

OTHER BOOKS BY MORTON WALKER, D.P.M.*

THE TOXIC METALS SYNDROME: How Heavy Metal Poisonings Affect Your Brain (Avery Publishing Group, Inc., 1994, paperback, $17)

SEXUAL NUTRITION (Avery Publishing Group, Inc., 1994, paperback, $13)

SMART NUTRIENTS: A Guide to Nutrients that Can Enhance Intelligence and Reverse Senility (Avery Publishing Group, Inc., 1994, paperback, $13)

FOODS FOR FABULOUS SEX: Natural sexual Nutrients to Trigger Passion, Heighten Response, Improve Performance & Overcome Dysfunction (The Magni Co., 1993, paperback, $20)

THE COMPLETE FOOT BOOK: First Aid for Your Feet (Avery Publishing Group, 1992, paperback, $16)

INSTANT PAIN RELIEF: A Proven New Method for the Immediate Resolution of Chronic Pain and Dysfunction (Biological Publications, 1991, paperback, $16)

THE POWER OF COLOR: The Art & Science of Making Colors Work for You (Avery Publishing Group, Inc., 1991, $13)

PAIN, PAIN GO AWAY: Free Yourself from Chronic Pain (Ishi Press International, 1990, paperback, $18)

THE HEALING POWERS OF ELDERBERRY INTERNAL CLEANSING (NEW WAY OF LIFE, INC., 1989, paperback, $8)

SEXUAL NUTRITION: The Lover's Diet (Zebra Books, 1988, paperback, $9)

THE HEALING POWERS OF GARLIC (NEW WAY OF LIFE, INC., 1988, paperback, $7)

THE YEAST SYNDROME: How to Help Your Doctor Identify and Treat the Real Cause of Your Yeast-Related Illness (Bantam Books, 1986, paperback, $9)

COPING WITH CANCER: How to Fight Malignancy with Alternative Methods of Healing (Devin-Adair, 1985, paperback, $7)

SECRETS OF LONG LIFE: How to Remain Healthy, Sexual and Actively Working Well Beyond 100 Years (Devin-Adair, 1983, paperback, $7)

HOW NOT TO HAVE A HEART ATTACK (Franklin-Watts, 1980, paperback, $13)

THINK AND GROW THIN: Lose Weight, Gain Health, Live Longer (Arco Publishing, 1978, paperback, $7)

SPORT DIVING: THE INSTRUCTIONAL GUIDE TO SKIN & SCUBA (Henry Regnery Co., 1977, paperback, $18)

YOUR GUIDE TO FOOT HEALTH: The Book for Total Comfort for Your Feet (Arco Publishing, Inc., 1973, paperback, $7)

* These, and 40 other books by Dr. Walker, are available from NEW WAY OF LIFE, INC., 484 High Ridge Road, Stamford, Connecticut 06905-3020, U.S.A.

THE CHELATION ANSWER

How to prevent hardening of the arteries and rejuvenate your cardiovascular system

Morton Walker, D.P.M.
in consultation with
Garry Gordon, M.D.
Introduction by
William Campbell Douglass, M.D.

TO

Joan Walker,
who has sacrificed much
to further the cause
of chelation therapy

Acknowledgment is made to Michael B. Schachter, M.D., who read the text before publication and offered suggestions for medical exactness.

Library of Congress Catalog Card Number
93-86538

ISBN 0-9626646-7-7

First Paperbound Edition 1994

For information regarding this book, call or write:
Second Opinion Publishing Inc.
1350 Center Drive, Suite 100
Dunwoody, Georgia 30338
(800) 728-2288

Contents

Please Note:

This book is for informational purposes only. It does not substitute for the medical advice and supervision of your personal physician. No medical therapy, including chelation therapy, should be undertaken except under the direct supervision of a responsible practicing physician. The publisher, author, and contributors accept no responsibility or liability for any damage or loss that may be incurred as a result of the use or application of any information included in this book.

Preface

As a medical journalist, I was elated when I published the first consumer book ever to be written on chelation therapy. That book, *Chelation Therapy: How to Prevent or Reverse Hardening of the Arteries*, furnished medical consumers living in western industrialized nations full disclosure of a truly preventive medical procedure.

In fact, it popularized this marvelous treatment and eventually made me famous as a literary specialist in researching and writing about orthomolecular nutrition, holistic medicine, and alternative methods of healing. Indeed, the newly established Office of Alternative Medicine, National Institutes of Health came into being precisely because of just such books. And the American people were awakened by public information about medical alternatives which surpass in effectiveness the standard methods of conventional medicine as practiced during past years. I owe a great deal of this success to the hard-working people at the American College for the Advancement in Medicine (ACAM).

ACAM (then called the American Academy of Medical Preventics or AAMP) was also instrumental in getting me to write a new book — one that contained the full protocol for the administration of chelation therapy giving ACAM's chelating doctors a reference textbook. Knowing that there's only a limited market for professional textbooks, the publisher and I set out to produce a highly readable technical book which would be enjoyed by patients as well as physicians. In consultation with that brilliant physician, Garry F. Gordon, M.D., the resulting book was *The Chelation Answer: How to Prevent Hardening of the Arteries and Rejuvenate Your Cardiovascular System.*

This book is unlike any ever written on chelation therapy. Nothing else published comes near to being packed with so much relevant information. Every word on these pages is as true today as it was 10 years ago. In fact, the last 10 years have been spent testing and confirming the method of administration, conditions responding to treatment, lack of side effects, and more. This book is the only place that a lay person will find the complete chelation therapy protocol which is followed by chelating physicians who are members of ACAM.

A literate individual studying this text will know nearly as much about the treatment as do many chelating physicians. Certainly, the medical consumer who reads *The Chelation Answer* will know tremendously more about chelation therapy than does the conventionally practicing physician who follows the unbending and tunnel-visioned party line of the American Medical Association. There is no cookbook medicine here!

Along with the protocol, I have presented twelve signals of cardiac involvement which, when known, will allow persons at risk to take defensive actions, preserve their lives, and improve their quality of living. I advise you about the heavy metals that chelation therapy removes from the body. And I discuss symptoms, occupations, products, and other sources of poisonings from mercury, lead, aluminum, and cadmium.

Did you know that chelation is a highly effective treatment for cancer? There's much elaboration on that subject as well.

Personally, I've received chelation therapy more than eighty times. My practice is to take a few booster treatments every year or two — not for any health problem but just to live as close to my genetically coded 120 years as I can. Besides the United States, I've received chelation therapy in England, New Zealand, Denmark, Germany, Australia, Canada, and Mexico. At my recommendation my mother has undergone a series of chelation treatments, and so has my wife.

Assuredly, my book tells the whole story of politics and greed as represented by the rapacious health insurance industry, organized narrow-minded American medicine, corrupting influences of the pharmaceutical industry, and their Gestapo-like bed partner, the U.S. Food and Drug Administration.

What an interesting story chelation therapy has turned into. It's so compelling and ever-changing that I must produce a new topical book about every three years. Even so, of the 55 books that I've authored to date, *The Chelation Answer* is among the best that I've written. Moreover, it is one of which I am exceedingly proud.

Since this book was first published, the protocol contained within has saved countless lives by eliminating the need for heart surgery, preventing amputations, detoxifying metallic poisons, defusing digitalis overdoses, removing hypercalcemia, and correcting heart arrhythmias. The protocol additionally has relieved or even reversed the symptoms of arthritis, macular degeneration, diabetes, hypertension, scleroderma, and two dozen other serious health problems. Read the book and see!

Morton Walker, D.P.M.
Stamford, Connecticut

Foreword

When Dr. Gordon and Dr. Walker submitted this manuscript to me and invited me to write a foreword, my first response was to ask, "Why me? I've never chelated anyone."

"That's just why we think you can be objective. Your writings certainly show a willingness to look into new ideas, with the welfare of the patient and not of the doctor as the focal point," they replied.

And so I did look into chelation therapy and have tried to be objective. I studied the arguments, pro and con. I studied the reports of chelating physicians and of orthodox physicians.

Having done so, I have concluded that chelation therapy is a safe, useful, extremely promising therapy for many cardiovascular conditions, albeit somewhat expensive and time-consuming. Further, I am inclinded to believe its value may even extend beyond its cardiovascular application to include prevention of cancer and immune disorders and the retardation of aging.

And now I invite you, with the help of this book, to make a similar study. I believe you will find all the pertinent information within its covers.

Chances are that you will read this book and become enthused about the possibilities of chelation therapy helping you (or a family member) with some serious medical problem. You will then proceed to ask your personal physician or consultant in cardiology, neurology, or the like, if he does not agree that chelation therapy should be started.

And chances are he will reject your suggestion out of hand. "No way," he may say. "Chelation is dangerous, or it's quackery, or it's an ineffective waste of money."

I always point out to my patients that whatever befalls them is, in reality, a matter of the choices they make. Look how that axiom applies here. Your doctor's negative response calls for one of two choices on your part. You may either question the veracity of the chelation story, documented as it is down to the last detail, or you may question the judgment of your physician.

Well, if your choice is not to question your doctor's judgment, you may well be making a grievous error. After all, you will have read the same book that I did. You will have seen that it is exhaustively complete and meticulously documented. After you've finished reading this book, I will ask you a question: Do you see any basis for a well-informed person arriving at the conclusion that chelation therapy has no value? Of course not. It should be obvious to you that any doctor who maintains that chelation therapy is not a "treatment of choice" in most cardiovascular disease is maintaining an untenable position.

It's worth a few moments to find out why your doctor has placed himself in a position that cannot be justified on scientific, ethical, or moral grounds.

Is it because he is ill-informed or misinformed? Let's hope so, because the "cure" for a lack of information is information. When a book as comprehensive as this one exists, it can serve to dispel any lack of knowledge on the subject. Tell you doctor to read it carefully.

Or does it represent the follow-the-leader mentality adopted by most physicians? Despite their seeming financial independence and freedom from the constraints of being mired in some corporate hierarchy, most doctors seem pathetically unwilling to "leave the pack," as if motivated more by the fear of departing from established protocol than by the desire to help their patients get well.

Or is it a manifestation of the closed-mindedness that provides a comfort to so many of our physicians? I personally know many doctors whose mentality runs thus: "If it were so good, don't you think we'd all be doing it?"

Perhaps the greatest blame rests upon the naiveté of most doctors in failing to recognize the economic self-interest that motivates their leadership. One source of these financial considerations is derived from the power of the drug industry, which depends on the prevalent philosophy among health professionals that administering these metabolic poisons (which drugs are) can somehow enhance health better than can the natural enabling agents found in our diet and in our bodies. The difficulty here is that the drug interests dare not give credibility to the chelating doctors, most of whom are strongly schooled in the pharmacologic possibilities in nutritional supplements. They know that when the truth gets out, as it will when the chelating doctors are heard, physicians' prescribing practices should be in for a major modification.

Another source of economic pressure is derived from the economics of hospital management, where expensive equipment for open heart surgery, angiocardiography, and the like represents major fiscal investment that must be amortized by extensive patient utilization.

Perhaps the major economic pressure opposing chelation is mediated by the insurance carriers, who simply fear an unexpected negative cash flow on insurance they have already underwritten.

But whether it be through naiveté, arrogance, or ignorance, the failure of your negatively disposed physician to recognize what is obviously in your best medical interests indicates simply that he does not deserve to be your doctor.

In this world where medical controversies abound, remember this truth: An expert with no experience in a subject is not an expert in that subject. This means that the consultant cardiologist who knows well the consequences of vascular surgery and pharmacology but who is not conversant with the pros and cons of chelation therapy, is only an expert in what he has experienced and is *not* an expert in *all* vascular therapies. He is therefore, unable to render a decision in any case in which chelation therapy must be considered. Why, then, should you heed his advice?

Well, I, for one, am sick and tired of the arrogant pronouncements of these inexpert "experts." By their mouthings, controversy is created where none deserves to exist. Among the many physicians who have firsthand experience with chelation, *there is very little controversy*. I have yet to meet a chelating physician who is not enthused about this form of treatment and who does not urge me to follow suit.

So, I am hereby announcing that all of you, as well as this very book, have convinced me that it simply is not fair to my patients not to offer them chelation therapy. And so, by the time this book has gone to press, you will be able to add my name to the roster of chelating physicians.

Robert C. Atkins, M.D.
New York

INTRODUCTION

Chelation Works!

As with so many things in life, I became a chelation advocate (and Public Enemy Number One with my medical society) after an encounter with a dying man at a small hospital in Port Charlotte, Florida. I had switched nights on duty in the emergency facility as a favor to one of my colleagues. At eight o'clock in the evening, a 45-year-old man named Al was brought in with severe chest pain. As I was interviewing him, he had a seizure. We were well-trained in cardiac emergencies — this was Retirementville, U.S.A., where many people died a cardiac death every day — we really knew our stuff, and were proud of it. Chest pain plus convulsion equals ventricular fibrillation — and a rapid death unless immediate action is taken.

Al had chosen the perfect place to have a massive heart attack. Because of his complaint, we had immediately started an intravenous drip when he came in "just in case." The electrodes had been placed on his chest in the ambulance so we had immediate confirmation of the diagnosis when he convulsed: ventricular fibrillation. The electrical paddles, used to convert the heart to a normal rhythm, were at my side, a breathing tube was ready, and the oxygen was flowing. Everything required to bring Al back from the brink of death was in place and immediately applied. The whole life-saving procedure — defibrillation, intubation, respiration, the administering of certain drugs — took four of us about 45 seconds. Boy, was he lucky to be at the right place at the right time.

His wife, Marty, witnessed the entire drama and, after he recovered, told him about it.

About a month later, after Al had been released from the hospital, he brought a tabloid newspaper into the emergency department and asked me about a "new treatment" called chelation therapy, prominently featured in the tabloid. He was considering taking this miracle therapy. It was said to "dissolve away" the rust and crud in the arteries and thus enable the patient to avoid coronary by-pass surgery. He had suffered one of those, and that was more than enough.

I patiently explained to Al, an uneducated, very smart, and street-wise Italian restaurateur who could afford any treatment he wished, that this was a *tabloid* article and therefore wasn't worth the yellow paper it was written on.

As I said, Al was street-wise; I wasn't: "Listen, Doc, just because this ain't in the AMA whata-you-call-it journal don't mean its a lie. I know it ain't the truth, necessarily, but do *you* know enough about it to say its bull----?" He really had me. It was difficult for him to talk to me like that, he told me later after we had become good friends. He had an immense respect for doctors but he'd jump on anything, or anybody, where his health was concerned — Al was a survivor.

Nevertheless, Al had piqued my interest and after a number of false starts I found, to my surprise, an organization of doctors who advocated chelation therapy. I studied their research information, listened to their case histories, went to their meetings, and have been an enthusiastic supporter and practitioner of chelation therapy ever since. I have treated hundreds of patients, starting with Al, and have seen some truly remarkable results.

The story of Al has a sad, and for me, frustrating ending. He did so well on chelation that he began to feel invulnerable — so much so that he went back to smoking and working 14 hours a day. In spite of this abuse of himself, he continued his lifestyle completely free of pain or other symptoms of heart disease for five years. He felt so great that he decided to stop the treatments in spite of Marty's urging and my warnings.

Six months after stopping the treatments he had a massive heart attack and died in the same hospital where we met. *The doctors blamed the chelation therapy and me for his death.* Marty's retort to them was even more unprintable than Al's would have been. She could have cooperated with these doctors, ruined my career and made a bundle — "pretty young girl's husband dies at the hand of a quack" — the local medical society said so — case closed; it would have been easy. I tell you this story so you can appreciate how tough it is for a doctor to practice outside the mainstream. I'm surprised we aren't *all* in jail.

CHELATION'S UPHILL BATTLE

Chelation therapy is the treatment of hardening of the arteries with a synthetic amino acid administered into a vein. The fact that it has been almost ignored to death by conventional medicine is a sad commentary on our medical industry. If it were to be accepted as the valid modality it is, the pharmaceutical industry, the doctor industry, and the surgical industry would suffer grievous financial harm — chelation therapy just isn't lucrative enough.

But, why the insurance companies would be opposed to a therapy that prolongs life is beyond me. Chelation therapy, by prolonging health, and perhaps life, would cause *a positive cash flow* to the insurance companies — the longer you live, the longer you pay premiums and defer their cost of "insuring" your life. Why would they refuse to pay for a treatment that increases profits? The only answer I can come up with is the don't-rock-the-boat syndrome: We're making money — don't change anything.

But ignoring this rapidly growing technology didn't work; it was gaining ground quickly because of its obvious efficacy. So then the horror stories began to appear in the lay and scientific press: Chelation can destroy the kidneys (true if misused); chelation can dissolve your bones causing osteoporosis (not true — it may even *help* osteoporosis); it doesn't work (haven't heard that one recently because it obviously does work).

The most effective and invidious ploy against chelation has been the blaming of terminal patients' deaths on the treatment. Here's a typical scenario: A 70-year-old man has had two "bye-pass" operations. He is in such bad shape that he can't go to the bathroom without help. His brain is obtunded and he is considering suicide. He has a chronic bladder infection and possibly an infection of a heart valve secondary to the operation. He had a stroke following surgery — not an uncommon occurrence — and cannot use his right hand. And certainly not least important, he is in constant pain from the huge incision the surgeons had to make through his chest in order to stomp on his heart.

The family, realizing that he is worse off than before the surgery, takes him to a doctor specializing in chelation therapy. They are not stupid suckers; they are not uneducated peasants; the facts are quite simple: Dad went in for coronary by-pass surgery; they spent $100,000 of the insurance company's and their own money; Dad's much worse off. It's as simple as that. Modern medicine has made an invalid out of a loving husband, father, and grandfather. But the chelation therapist, starting out with a massive handicap, gets the blame from the doctors when the patient dies.

I forget who said it: "It's better to die than go against the faculty of medicine." But people are now turning against the "faculty of medicine" and are seeking other avenues of therapy — in droves. And chelation is an alternative to drugs and surgery that has been rapidly gaining the confidence of people the world over.

THE SCIENTIFIC SIDE

Chelation refers to the ability of certain chemicals to bind with calcium, iron, and other metals, and remove them from the body. We know there is a lot of calcium in those plaques that appear in the arteries of patients with "arteriosclerosis." Presumably, if calcium could be "clawed out" of a constricted vessel ("chelation" comes from the Greek word, *chele*, meaning claw) then the plaque could be slowly dissolved away. This is probably *part* of the explanation for how chelation works, but there is a lot more to it than that. In fact, the more we learn, the more mysterious it gets.

The primary chelating agent used in clinical practice is called EDTA, ethylene diamine tetraacetic acid. EDTA is a synthetic amino acid first produced in Germany in 1930. It was used as a preservative in cloth, and later as a stabilizer for food, such as mayonnaise (check your Hellman's label and you'll see that you have been eating EDTA all your life). In the 1950s, Dr. Norman E. Clarke, chairman of the research department, Providence Hospital, Detroit, Michigan, began research on the effects of EDTA chelation therapy on cardiovascular disease.

Dr. Clarke's motivation for investigating the possibility that EDTA would do in the human body what it did for mayonnaise, a rather preposterous idea on the face of it, was based on the hopelessness he felt after 30 years of treating arteriosclerotic heart disease with no positive results: "I knew, having been in cardiology quite a number of years, that arteriosclerotic vascular disease was a helpless, hopeless situation for the cardiologist." Few cardiologists, then or now, have the courage to admit they do very little to help their patients.

Dr. Clarke gave some historic testimony before the Scientific Board of the California Medical Society in 1976 that few doctors are aware of. He reported that he had personally administered 120,000 infusions of EDTA chelation and never saw "any serious toxicity whatsoever. I've seen only benefits."

The first dramatic results seen by Dr. Clarke were in patients with blocked arteries as a result of diabetic vascular disease. He reported cases of gangrene of the toes, due to blockage of the arteries, that were saved from amputation of the foot by intravenous chelation therapy. This dramatic reversal of a "surgical problem" by the use of a medication had never been achieved before. Surgeons were not enthusiastic, however, and, strangely, neither were most cardiologists. Almost 20 years after Dr. Clarke's remarkable testimony, they're *still* not enthusiastic.

He next reported on the use of chelation in elderly patients with senility due to arterial blockage in the brain. He had similar dramatic results and remarked to the scientific board: "After all these years and after all that experience, I am just as certain as can be that EDTA chelation therapy is the best treatment that has ever been brought out for occlusive vascular disease."

I would like to emphasize that Dr. Clarke was not reporting on Alzheimer's disease, another matter entirely, involving a younger population. Alzheimer's generally affects people in their late 50s and early 60s, who lose their minds as though they were much older. It's basically an early senility, often accompanied with frank psychosis and eventual violence — a terrible fate, and one in which chelation therapy has been, in most cases, disappointing. Because of the ability of chelation to bind and eliminate toxic metals from the body, the therapy *might* be useful as a preventive, but there is no way to prove that.

Prevention is almost always impossible to prove, but I still heartily recommend chelation for anyone over 60 years of age, whether they have any symptoms of disease or not. Norman Clarke chelated himself well into his 90s, and he remained active. One case proves nothing, but a solid reason for preventive chelation therapy is this: *There is compelling evidence that chelation can prevent cancer.*

In 1972, a Swiss report indicated a preventive effect against cancer as a result of chelation therapy. The research had been done on the use of chelation in the treatment of lead poisoning but the follow up studies on those treated revealed that the participants in the lead treatment study had a 600 percent decrease in the incidence of cancer as compared to people in the same community not given chelation.

It is not generally known, even by most doctors, that almost all chemotherapy procedures used in the treatment of cancer involve some form of chelation. Most antibiotics work through chelation, and even aspirin is a chelating agent. So if your doctor says he doesn't believe chelation is a valid therapy, tell him he doesn't know what he is doing, from the biochemical point of view, and to go back to his chemistry books. (Be sure to have another doctor lined up first.)

The one area where chelation is accepted is in the treatment of heavy metal poisoning — especially lead. They can't deny it works there. But it is also effective in the removal of iron, a far more serious problem in the general population than lead poisoning. Americans in general are overloaded with iron, primarily due to a misguided food industry that adds iron to foods as a "fortification," because the experts at the FDA say it's good for us. Well it isn't good for *most* of us — it's *bad* for us. Excessive iron leads to hardening of the arteries and iron is probably the most important toxin in our foods — not cholesterol, not fat, not salt — *iron.*

The removal of chelated iron may be the most important action of chelation because excess iron not only causes arteriosclerosis but cancer and infection. So there are many reasons why you should take prophylactic chelation therapy.

Patients who come in for chelation for their vascular disease are often pleasantly surprised to note their arthritis improved. The reason for this is not known. It may be due to removal of iron, or calcium, or both, or it may be due to the improvement of the circulation to the joint — or all, or none, of the above. But, for whatever reason, the improvement in arthritis is often dramatic.

In my experience, the first sign of improvement is often seen in the skin. Many patients remark on the improvement in their skin color, the disappearance of blemishes and better skin turgor.

With heart patients, the first sign of improvement is usually an increase in exercise tolerance. I have had patients who had difficulty crossing the living room who, after 20 treatments, could walk to the mail box, a hundred yards down the path, with no difficulty. (Sometimes my bill would be in there. It's easier to collect when the patient can walk to the mail box and he couldn't before.)

The procedure is relatively simple. The doctor will check you out for kidney malfunction, heart failure, and any other condition that would indicate the need for a modified, more cautious approach. There are few patients who cannot tolerate the treatment. You will sit in a reclining chair for two to three hours taking an intravenous drip. The most common side effect is boredom, so bring an interesting book.

You should commit yourself to 30 treatments. Don't expect miracles with the first infusion — you didn't get in this terrible shape overnight.

There are other methods of clearing the arteries which are complementary to chelation. Hydrogen peroxide, also given intravenously, has a chelating effect, although the mechanism by which it works is different. And we now have a third therapeutic weapon, photoluminescence, which treats the blood with light and has a mechanism of action similar to hydrogen peroxide — a marked increase in the oxygenation of the body's organs and tissues. In the chelation protocol of the near future, all three of these treatment modalities will be used together for the alleviation of many diseases.

William Campbell Douglass, M.D.
Turku, Finland

CHAPTER ONE

You Can Prevent Hardening of the Arteries and Rejuvenate Your Cardiovascular System

What if you heard that your case of severe hardening of the arteries need not lead to open heart surgery or heart attack?

Suppose someone told you that your father's leg, scheduled to be amputated because of diabetic gangrene, did not have to come off?

How would you react to learning that your senile old mother, vegetating in a nursing home, could be restored to the active, alert, productive person she once was?

What will you do if macular degeneration has taken your daughter's eyesight, and you've just found out that her blindness can be reversed by a series of intravenous injections?

Suppose a stroke recently left your spouse paralyzed and speechless; would you be interested in a safe, nonsurgical, medical procedure to remove these tragic symptoms?

Having such startling revelations laid before you, more than likely you'll become amazed and perplexed, right? Maybe even be dubious! You might wonder why you hadn't heard of such dramatic treatment before.

Will it be a surprise if your local doctor says he isn't at all acquainted with this procedure called *chelation therapy* and possibly would be reluctant to prescribe it even if he knew something about it?

Well, if you plan to read further, you'll be exposing yourself to an incredible medical education. The knowledge will probably have you responding positively and quickly to prevent further deterioration of your own or someone else's cardiovascular system. You may even feel resentment that the information has been kept from you until now.

Chelation therapy, the way it's employed in the protocol of the American Academy of Medical Preventics, doctors joined together in a medical specialty organization devoted to rejuvenation of the cardiovascular system, should be made available to anyone who is suffering from an impairment of the flow of blood to the head, limbs, heart, and other internal organs. I make this statement in the face of the California Medical Association's taking a political stand against holistic alternatives by saying that the treatment is not yet proven and must be considered experimental. As recently as May 29, 1981, *The Medical Letter on Drugs and Therapeutics* reported, "There is no acceptable evidence that chelation therapy with EDTA is effective in the treatment of atherosclerosis, and the adverse effects of this drug can be lethal." Some doctors have even told their patients that intravenous chelation is useless quackery or fraud. Although the American Medical Association has declared that the treatment's status and worth for reversing hardening of the arteries "must be regarded with skepticism," the AMA still acknowledges that physicians are perfectly free to use EDTA chelation therapy if in their best judgment they believe it would be the correct thing to do for their patients.

The story of chelation therapy will be presented here as I know it from statements made by the courageous physicians who provide the treatment, frequently against tremendous peer pressure not to offer it. Also submitted will be case histories of people I have interviewed. They will tell of their experiences with recovery from life-threatening conditions by taking a series of intravenous infusions of the EDTA chemical.

As a person possibly at risk of being one of the millions suffering each year from heart attack, stroke, high blood pressure, and other degenerations of the circulatory system, you must have this information to make important health care choices. Is the chelation alternative to surgery, drugs, and other forms of establishment medicine right for you? All of us must eventually come to such a decision.

George W. Frankel, M.D., of Long Beach, California, was one of those forced to make his own life-saving decision. While golfing under the Southern California sun in December 1971, Dr. Frankel got the shock of his life when he was suddenly seized by severe chest pains. His golfing partners helped him return to the clubhouse, and after resting, Dr. Frankel visited his cardiologist.

Following electrocardiograms and treadmill tests, a coronary angiogram (a heart catheterization) was performed at the hospital. Dr. Frankel was told by the cardiologist and a chest surgeon that this angiogram revealed obstructing plaques (tumorlike bumps) in the left main coronary artery. They recommended a triple coronary artery bypass. This operation must be done as soon as possible, the specialists said, in order to avoid heart attack and possibly sudden death.

Coronary artery bypass surgery involves removing one of the major veins from the patient's thigh and patching it in to bypass the obstructed arteries feeding the heart. Such surgery entails several serious hazards. Under the best conditions, there is a 1 to 4 percent possibility of a heart attack while on the operating table; under the worst conditions, a 10 to 17 percent rate of death in the operating room occurs. There can be further deterioration of the arteries bypassed, with a 20 to 30 percent possibility of blockage of the new graft within two years; and there is the possibility of the reappearance, even within a few months, of similar chest pains, with little or no chance of relief by additional surgery. Although the eventual overall mortality of this operation at this writing averages approximately 5 to 12 percent, depending on the skill of the surgeon and the severity of the condition, morbidity—better known as the complications—still occurs in another 10 to 15 percent of the cases done. These percentages figure out to about one in twenty-five coronary artery bypass patients dying during the operation; one in seven will suffer complications after the surgery is over.

Physicians experience fear of life-threatening procedures, as does anybody else, and this ear, nose, and throat specialist was no exception. Dr. Frankel, chief of otolaryngology at two hospitals, was not anxious to face major surgery on his heart, but there seemed to be no alternative. His doctors offered nothing else. Consequently, he agreed to the triple bypass.

While waiting for his surgery to be scheduled, Dr. Frankel discussed his condition with many of his colleagues and friends. One friend detailed the case history of someone with a similar condition. The individual underwent a treatment, previously unknown to Dr. Frankel, called *chelation*, a medical therapy that was thought to reduce the amount of calcium in the obstructing plaques of the coronary arteries and in other areas of the vascular system.

Dr. Frankel, eager to try any reasonable, safe, and painless avenue that might avoid the very real potential of death offered by open heart surgery, began to search for clinical literature on the subject at the Los Angeles

County Medical Association Library. He was amazed to find numerous medical journal articles on chelation therapy. After careful study of these many references in the *Journal of the American Medical Association, The Lancet, Angiology, Southern Medical Journal, American Journal of Cardiology,* and dozens of other medical journals from around the world, the patient ventured on what he hoped would be a journey to save his life. He entered an Alabama hospital that was run by H. Ray Evers, M.D.

"I saw people come in with diabetic ulcerative lesions and gangrenous lesions that cleared up in a matter of ten or twelve days, and I could not believe my eyes," said Dr. Frankel. "I saw one patient who was admitted after being told elsewhere that his leg would have to be amputated because of gangrene, and daily, after chelation therapy, I watched with my mouth agape as the leg came back to normal. I acted as a sort of assistant to Dr. Evers by making rounds with him every day. He was a very determined man who worked twenty of every twenty-four hours."

There was something else that Dr. Frankel could hardly believe. Prior to treatment, anginal pains had seared his chest when he walked only ten or twelve steps. After he received only ten chelations, the pain disappeared entirely. He decided to put off having the heart surgery indefinitely. In the event he needed an operation later, the patient knew, his chance of survival was considerably enhanced by the remarkable clinical improvement he had experienced from his chelation treatments.

The physician-patient has since given himself over 100 more chelations at home, with the help of a nurse-anesthetist for the intravenous injections—a total of at least 128 treatments to date. "And I want to tell you that I have not had an angina since early January 1972," says Dr. Frankel. "I carry a full work load. I perform approximately ten to fifteen surgeries every week, and these are microsurgeries of the ear. I carry a full practice. I play golf. I swim twenty laps in my pool every day, and I cannot speak with any but the greatest praise for the men who are attempting to make chelation an accepted form of therapy."

Of all the professions, the practice of medicine is more prone to produce coronary and blood vessel disease than any other. Doctors know this.

On the morning of June 9, 1974, after his usual fitful night's sleep, Richard E. Welch, M.D., a sixty-year-old general practitioner of Hesperia, California, arose, shaved, and stepped into the shower. As he soaped himself down, the physician began feeling nauseous and dizzy. Then he was struck with a severe headache. Dr. Welch immediately recognized these as dangerous symptoms, for he had consciously ignored other signals of something gone wrong in his body. He was experiencing insomnia, reversal of the day-night ratio of urinary output (nocturia), slight but persistent headaches, sluggish memory, high blood pressure, a general feeling of tension, transient weakness in a limb, and real or imagined impaired vision.

Dr. Welch steadied himself and moved from the shower stall. He resolved to visit his internist colleague that very day for a physical examination. He promptly collapsed onto the bathroom floor.

Mrs. Welch helped her husband back to the bed, took his blood pressure, and recorded a finding of 230/190. The man had sustained a severe cerebral vascular insufficiency, or what is know as a TIA (transient ischemic attack). He came very close to having a complete starvation or strangulation of a part of the brain that medicine commonly labels as *stroke* or *cerebrovascular accident*.

Until this crisis occurred, Dr. Welch admits, he lived the same lifestyle that kills most Americans: a stressful daily existence, high caloric intake, too much fat in foods, lack of regular exercise, as well as some of the other very serious deficiencies and imbalances so common in the average American diet. The vast majority of us therefore die in what could be the prime of life, which is the fourth to the sixth decades.

"I recently took a course of chelation with two friends," said Dr. Welch. "Together the three of us felt remarkable improvement in our circulation by the time the course of treatment was over.

"It was not until I had received chelation therapy that my blood pressure came down to a normal level," he said, "even though I was on strong hypertensive drugs. The additional dividends of the treatment were more energy, alertness, no headache, less nocturia (urinating at night), and reactivation of the libido."

Dr. Welch pointed out that he has reviewed extensive clinical data, involving thousands of cases that are beautifully documented by special studies of circulation such as thermograms (which measure body heat, a function of circulation) before and after chelation therapy. These studies prove the value of this form of treatment.

Patients with any of the various signs of hardening of the arteries such as cold extremities, poor exercise tolerance, nocturnal leg cramps, intermittent claudication (leg pain that starts and stops with walking), or even pains of angina pectoris, may show spectacular improvement from their series of chelation treatments. A large body of clinical proof is available immediately, which the American Academy of Medical Preventics is anxious to supply to any interested medical personnel, governmental officials, or even ordinary people. The improvements are demonstrated with clarity and detail by traditional, sensitive, and highly accurate, painless, noninvasive, and easily reproduced tests of circulation available in most medical centers and the offices of many preventive medicine doctors throughout the United States. These noninvasive tests, which will be described further in Chapter 4, include the ultrasound Doppler, plethysmography, thermography, the treadmill electrocardiogram, and radioisotope blood flow studies performed on the heart and brain.

On March 31, 1974, Lester Tavel, M.D., D.O., Ph.D., now of Bradenton, Florida, suffered a heart attack and immediately developed a paroxysmal atrial tachycardia (speeding heart with little pumping power) with a pulse rate running at 200 beats per minute. It took three electric shocks of 25, 50, and 100 volts to cardiovert (electrically countershock) the victim back to a near-normal heart rhythm. An X-ray examination of his chest showed the patient's heart was dangerously enlarged, filling practically the whole chest cavity. He was brought to a hospital's intensive care unit, where he remained. Because of the risks that went with open heart surgery, such an operation was not even considered by the patient, his attending physicians, or his family.

As soon as Dr. Tavel could be moved safely, his wife flew with him to the office of his friend Harold W. Harper, M.D., a chelating physician in Los Angeles. Because her husband was unable to take more than three or four steps without experiencing dreadful shortness of breath, Mrs. Tavel pushed him in a wheelchair. The patient's ankles were badly swollen as a result of his failing heart.

Dr. Harper set up intravenous chelation in an apartment-hotel suite he had made ready, since most establishment hospitals in the United States won't permit chelation therapy for hardening of the arteries or heart disease to be given using their facilities. At the end of three weeks, during which he received fifteen chelation treatments, the patient's heart began to return to its usual size. In another week, upon X-ray examination, it again appeared completely normal.

"You know, Los Angeles has a lot of streets with forty-five degree inclines," Dr. Tavel later said. "Well, by the time I received thirty chelations in about six weeks, I was walking up the hill in front of my hotel. I walked up steps, too. My resting pulse rate was 84 beats per minute, and by the time I reached the top of the hill, it increased to 110. Then it dropped back to 84 upon my resting within a minute afterwards."

"What Lester didn't know at the time," said Dr. Harper, "is that his electrocardiogram [EKG] and enzyme studies indicated he had suffered an acute myocardial infarction [a local area of death] in the heart muscle. I began to administer chelation therapy as soon as my emergency medical workup for him was complete. After about the first five treatment days his shortness of breath began to go away. The ankle swelling was down. He was able to eat again. His color changed from pasty white to something near his natural ruddiness.

"I took a second X-ray series after the patient had received about ten chelations. His heart size showed close to normal, but not quite. Lester's enzyme studies and blood sugar had returned to normal, though, and his EKG returned to normal within a two-week period.

"Follow-through at the end of thirty treatment days showed my patient's heart size comparable to the way it had been when I saw him the previous November. Comparison X-ray films attested to the heart sizes as being exactly the same. There were no fluid levels in his lungs—no congestion—no edema. Lester was able to go home at the end of six weeks," concluded Dr. Harper.

"I checked myself quite cautiously," Dr. Tavel added. "I ran a heck of a lot of BUNs [blood-urea-nitrogen tests for toxicity] and creatinines [urine tests] after I got home, and I didn't have any problems with those or any other toxic symptoms.

"I did notice many things about myself improve—my prior dyspnea [frequent rapid breathing] was relieved; my fatigue was relieved, my limbs were warmer," Dr. Tavel said. "I had a regrowth of hair on my legs, and I had an increased sex drive that my wife enjoyed. I gave myself ten more chelation treatments at home, and I've been taking six treatments a year since.

"About thirty days ago I had another chest X-ray that showed my heart size remaining normal. An independent group of cardiologists then evaluated my EKGs and rechecked about thirty-five of my laboratory tests, including the many heart enzyme tests and liver enzyme tests. All the diagnostic findings were back to normal, as if I never had experienced a heart attack," the physician said.

At the time of our interview, Dr. Tavel was engaged in an exceedingly active medical practice. It involved all the physical stress, emotional trauma, and long hours usually required in medicine. Nonetheless, since his heart attack in early 1974, he has vigorously worked sixty hours a week.

Each story you have read or will read in this book, I have double-checked for truth and accuracy. Dr. Tavel's story struck me as quite dramatic. His heart problem had him closer to death than anyone ever ought to be. I therefore took the opportunity to triple-check his story and sought out and interviewed someone who had worked for Dr. Tavel in Houston, Texas, back when he had experienced his near-fatal attack. I traced down and spoke with thirty-six-year-old Deborah Triche, a registered nurse employed by Dr. Tavel and his associate at the time, Dr. John Mohney.

Nurse Triche retold the story of Dr. Tavel's heart attack that I have described. She said, "I was the technician who recorded Dr. Tavel's EKGs and submitted them to our cardiological EKG service. The heart doctors on the service read the recordings and computerized their reports. Each time a patient has an EKG taken it is compared to the one taken of him previously.

"Well, after Dr. Tavel's chelation therapy was completed, I asked one of the cardiologists about the comparisons between his pre-therapy and post-therapy EKGs. The cardiologist said, 'You know Dr. Tavel's heart was severely damaged before the treatment, but now following chelation therapy it shows no aftereffects on the EKG at all.' I was amazed," said the nurse.

Currently, chelation therapy is administered to patients in Dr. Tavel's medical practice who need it. He is a prime proponent of the treatment.

When Harold W. Harper, M.D., was thirty-two years old and out of medical school just two years, his blood pressure stood at 220/150, and he weighed 358 pounds. Colleagues advised the new physician to leave the profession because the stress of medical practice was adding unduly to all his other heart attack risk factors. Cardiologists offered him only two more years of survival unless he altered his lifestyle and rid himself of some risk factors.

Being apprehensive about his well-established and irrevocable family history of hardening of the arteries, Dr. Harper took the precaution of carrying nitroglycerin tablets in his shaving kit wherever he traveled. He knew it was only a matter of time before he would be hit with a heart attack. His logic was that nitroglycerin might provide an immediate blood flow through the coronary arteries, until he could arrive at some emergency medical facility.

Heart attack struck the physician six years later, while he relaxed with friends aboard a houseboat on Lake Powell, located between Arizona and Utah. They had been fishing.

Dr. Harper's first sensation of something going wrong began with a headache "like little men inside my head with sledgehammers who were trying to get out," he later recalled. Another pain began just below his left breast at the rib margin. The pain rose slowly upward to his left shoulder, neck, jaw, and down the left arm. "This is it!" he told himself. "My time has come."

He rose to get the nitroglycerin from his shaving kit, fell flat on his face, and broke his nose. As he lay there, face down, semiconscious, unable to move or talk, his companions stared in shock. He attempted to move an index finger to let his stunned companions know he was alive, but could not do even that.

Dr. Harper's friends let him lie on the deck breathing fresh air until their party arrived at a landing. By fast powerboat they motored to a small airport and flew him for hospitalization near his home.

By the time the fishermen reached Los Angeles, paralysis had left the patient, and, at Dr. Harper's request, his friends took him to his office instead of to a hospital. There, using his own equipment, his medical staff gave the doctor an electrocardiogram, heart enzyme tests, and a full examination. The tests indicated that he was over the acute stage of the heart attack. He had undergone a serious disruption of the circulation through the coronary arteries.

Dr. Harper and the staff members decided that hospitalization would be of no benefit. He rested at home. The patient thereafter took time off from work for three days and then returned to medical practice on a modified work schedule that lasted for only three weeks; then he resumed his regular hours fulltime.

"A year later, I attended the annual meeting of the International Academy of Metabology in New York City. One of the lectures I tape-recorded was delivered by H. Ray Evers, M.D. He spoke on the use of chelation therapy for heart disease. This doctor from the little town of Cottonwood, Alabama, had a dramatic story to tell about chelation treatment. Frankly, I didn't pay much attention. It just sounded too miraculous. I am, after all, a scientifically trained person unable to believe the type of miracle clinical response that Dr. Evers was describing," Dr. Harper admitted.

"Yet, I couldn't push Evers's words out of my mind. That night in my hotel room I listened over again to the tape I had recorded. It made biochemical sense. I mulled it over for days. Back in Los Angeles I set a researcher to gathering all the literature about chelation therapy that could be found. The medical librarians pulled out everything in their stacks and photocopied hundreds of pages of information.

"During his lecture, Dr. Evers had invited any physician who wanted to learn more about chelation therapy to visit with him. After delaying two weeks, my curiosity got the better of me. I had my secretary telephone to tell the man I was flying in to find out more—to examine his patients' history charts, observe the people under care, make a decision about his results," said Harold Harper. "And that's what I did.

"I reviewed 150 history charts and made hospital rounds with Dr. Evers, beginning at 5:30 the next morning. He worked long hours. I noted the patients' response," said Dr. Harper. "People had come from all over the U.S.—Florida, Alaska, Oregon, Wyoming, Connecticut, Maryland, Montana—and from foreign countries also.

"People had heard of the treatment through relatives. Word can get around fast when some unusually effective reaction is taking place," Dr. Harper continued. "I saw patients who had arrived with Raynaud's disease, others with purple extremities, gangrenous lesions on the legs and feet, and the whole range of arteriosclerotic conditions. And I saw these conditions heal. They healed well! I was shocked! I decided to go home and perform the technique on myself. Dr. Evers explained to me how I could acquire the chelating solution and make the mixture. I became my own first patient for chelation therapy.

"I took serial EKGs on myself during the treatment. After just three chelation treatments my electrocardiograms returned to normal for the first time since I had entered medical school, almost twelve years before. My EKGs have remained normal since. My blood pressure became normal. My anginal pain went away and has not returned. That has been for ten years now," Dr. Harper told me.

WHAT IS THIS CHELATION THERAPY?

Chelation (pronounced *key-lay'-shun*) therapy consists of injections of a synthetic amino acid called ethylene diamine tetraacetic acid (abbreviated EDTA). When EDTA is introduced into your body through an intravenous infusion, the proteinlike material binds with or "chelates" certain minerals that are present in your bloodstream. Since calcium is the second most prevalent mineral or metal floating in the bloodstream, this chelating material has a profound effect on calcium metabolism and the mineral's availability. It locks onto the ionic calcium and removes it from the body, mostly through the urinary system.

Medical science now recognizes that a major component of circulatory impairment is spasm or constriction of the arteries.[1] Although the precise reason for this arterial spasm has not been entirely identified, blood vessel specialists now widely believe that the major problem involves some disturbance of calcium metabolism in the cardiovascular system.[2]

Sidney Alexander, M.D., Chief of Cardiology at the Lahey Clinic and Professor of Medicine at Harvard Medical School, lectured on this point to the American Academy of Medical Preventics during the Academy's Spring 1981 semiannual meeting. Dr. Alexander stated that medicine's recent recognition of calcium's role in blocking arterial blood flow is bringing about a pharmacological revolution in cardiovascular treatment.

Chelation therapy may soon begin to be recognized as the leader in that revolution. The U.S. Government and organized medicine have erroneously, until now, evaluated the treatment exclusively for its effect on arteriosclerosis when it actually should be appraised for improving the blood circulation through EDTA's anti-calcifying effects.

This new understanding about the underlying involvement of excess or abnormal accumulation of heart and blood vessel calcium has been tied to the aging process. It is known that if a person could avoid an excessive calcification of the cells in the arteries, he or she would avoid loss of arterial elasticity. The blood would circulate more effectively, bringing nutrition to all the body cells and taking away their waste products.

Research evidence from both living animals and human cadavers indicates that EDTA chelation therapy partially eliminates or reverses the formation of atherosclerotic plaque. Plaque is the pathological substance that tends to obstruct arterial blood flow in approximately 70 percent of all Americans by the time they reach age forty. While the reversal of plaque formation in human arteries has not been proven to the satisfaction of everyone in medicine, much documentation exists confirming that intravenous EDTA infusion does increase the circulation in most people having this chelation treatment.

It doesn't require that arterial blood channels must become wider or that plaque should become smaller. It's enough if the tiny collateral circulation around the clogging plaque is opened. Many capillaries and tiny arteries in the area of blood vessel obstruction will carry the blood around a clogged area if the artery is made more flexible and soft or if spasm is prevented; and spasm can't occur to as great an extent when excessive calcium levels in muscle cells are lowered.

EDTA chelation therapy is actually the first of an entirely new class of medicines known collectively as *calcium channel blockers.* They curb the flow of calcium and other minerals into the muscular coatings of the arteries and thus relieve chest pains, irregular heartbeats, and other symptoms that afflict an estimated twelve million Americans.

Of the four new calcium blockers, two, nifedipine and verapamil, have just come on the market, following approval by the Food and Drug Administration (FDA). Diltiazem and lidollazine are expected to be approved shortly. While the newest calcium blockers may relieve the excruciating chest pains caused by blood vessel *spasm*—only recently identified by medicine as cutting the heart's supply of blood and starving it of oxygen—EDTA chelation therapy does much more. Chelation actually helps prevent calcification and subsequent hardening of the arteries. According to world-famous researcher Johan Bjorksten, Ph.D., in his book, *Longevity: A Quest,* published July 1981, chelation therapy also holds great potential for extending your life expectancy by ten to twenty years.

Testing reveals that infusions of EDTA also help pull calcium from atherosclerotic plaque and the other areas of the body where it is abnormally deposited, such as in tendons, joints, and ligaments. But this treatment does not appear to significantly remove calcium from the bones and teeth, where the mineral serves a very useful purpose.

Diagnostic tests taken before and after chelation therapy reveal that areas of impaired circulation are often restored to normal by this newly recognized, efficient medical process. After it reduces muscular spasm of the artery and/or removes the clogging material from the arterial wall, a widened or unblocked, more flexible passage is left behind. Thus, blood can circulate without obstacles.

PATIENTS ARE NOT INFORMED FULLY

Chelation therapy has worked extremely well for more than three hundred thousand victims of hardening of the arteries who were lucky enough to learn of the treatment. A patient's blood pressure usually becomes normal, hands and feet grow warm with improved blood flow, kidney problems are averted, and the chances of stroke or heart attack are greatly reduced. Remarkable circulatory improvement takes place through all the blood vessels. Yet, for many reasons (mostly political), chelation therapy is seldom recommended by an orthodox physician as the treatment of choice for a person's impaired cardiovascular system.

As it stands now, someone who is suffering from hardening of the arteries generally receives grossly incomplete information about his exact circulatory condition and all of the different therapies available. The arteriosclerotic patient is definitely not told that chelation is an alternative therapy that he should carefully consider—perhaps because most physicians hardly know anything about the treatment themselves. In other instances, some physicians, believing they are well informed about chelation, may be basing their knowledge on hearsay or on totally misleading or incomplete or out-of-date information and not on current scientific knowledge. Although the average medical consumer may find it difficult to accept that wrongheaded thinking can exist in modern medicine since doctors are dealing with human life, nonetheless, medical information sources sometimes put out incomplete or misleading information, particularly when they feel threatened by a relatively simple, office-administrated treatment that could eliminate as much as 75 percent of the vascular surgery presently performed in the United States. As a result, some doctors may fail to inform a patient contemplating bypass surgery or limb amputation about chelation therapy as a viable alternative. They possibly never heard of it. Or they may mislead patients by erroneously claiming that chelation couldn't possibly help them. Furthermore, you should know at the very start that physicians practicing chelation therapy are in direct competition with chest physicians and vascular surgeons.

Of course, many orthodox doctors' reasons for not advocating this alternative approach are honorable. It might be that these physicians really do *not* believe the treatment would be effective. Legally, however, under present court decisions, doctors should inform patients of every alternative—even if they personally reject one or more of the choices. They are professionally responsible for knowing and presenting all the possible treatment alternatives.

Chelation therapy has been shown frequently to be remarkably effective in patients who were too sick to benefit from any other form of medical treatment, or were even too ill to undergo any type of surgical correction. Even so, patients may be led to believe that their condition is too advanced to consider a safe medical alternative such as chelation therapy. When confronted with chelation as a possible treatment, the individual doctor is likely to say: "If chelation therapy is so good, why isn't it being done by everyone in medicine?" Unfortunately, the answer is that plenty of physicians tend to be overly conservative. They are under peer pressure to offer well-established remedies. They are afraid to do that which is not routinely being employed generally in their community. In Chapter 5 I will discuss in detail the numerous reasons why your local physicians may not be providing chelation therapy for your community.

THE TREATMENT HAS BEEN USED A LONG TIME

Cardiovascular operations are prohibitively expensive and limited in their use, but these operations are terribly popular today. Surgery is restricted by the size and the accessibility of the blood vessels that can be operated upon. Tiny blood vessels cannot undergo surgical procedures, and many blood vessels are buried deeply within the tissues, such as in your brain. There they cannot easily be approached by the surgeon. In contrast, chelation therapy can readily be used to increase circulation anywhere in the entire cardiovascular system.

In contemporary chelation treatment there have been no documented fatalities as a result of its use when the established protocol is followed. As previously mentioned, death during or soon after bypass surgery is reported to average approximately 8.5 percent, depending on the experience of the heart surgeon and his staff and the kind of complications connected with the patient's condition. In surgical bypass the severity of the disease in the patient plays a major role. In fact, a study done by the National Institutes of Health that was reported to the medical community in March 1977 stated that no one should have a bypass operation except for the relief of intractable angina (chest pain). Yet nine out of ten of these same kinds of severe heart cases are reported entirely relieved of symptoms with chelation therapy.

Obviously, the risk factor is dramatically reduced with the chelation treatment as compared with the surgical procedure.

During and after treatment, most chelation patients are monitored carefully and encouraged to change their lifestyles to a healthier mode of living. They receive instructions on how to exercise daily and take high doses of vitamins E, C, A, and B-complex and minerals, all of which help the arteries remain healthy and flexible. Patients are advised to have periodic hair analysis to determine precisely what mineral supplements they should take to maintain the body's natural chemistry and optimal balance. They must stop smoking, too. But most of all it is the intravenous infusions of EDTA—usually from twenty to fifty of the three- to four-hour injections over a period of months—that return patients to an improved state of health and vitality.

EDTA chelation therapy has been used in the United States for hardening of the arteries since 1952. Even prior to this date, the treatment was employed for detoxification in cases of lead poisoning and even for radioactive metal toxicity, and it is approved for lead poisoning by the Food and Drug Administration.

At this writing more than 1800 scientific journal articles have been written about various aspects of EDTA chelation. The treatment has been shown to help in a wide variety of malfunctions and disabilities where the basic problem is an interference with the blood supply, especially as it relates to calcium metabolism. Its safety has been tested and proven by its use on hundreds of thousands of patients in about 3,300,000 separate intravenous feedings by over 1000 physicians for the past thirty years. EDTA chelation therapy has provided an overall average 82 percent success rate for improving circulation among all patients who have received an adequate injection series. These figures are reported from the clinical records of the more than 200 physician members of the American Academy of Medical Preventics, the doctors in the United States who have made chelation therapy for victims of hardening of the arteries a new specialty in medicine.

The unhappy truth is that too many dollars and too much time are required to bring chelation therapy to the majority of the American people. The plain and undeniable fact is that people are dying from hardening of the arteries in part because of an FDA induced "drug lag," substantially aggravated by a lack of acceptance of chelation therapy by organized medicine.

The nonacceptance is also partially due to an absence of significant pharmaceutical industry investment potential. EDTA is nonpatentable, in contrast to the newest calcium blockers, which the pharmaceutical industry is currently spending millions of dollars to develop and promote. Meanwhile, the most promising and clearly the most clinically tested calcium blocker of all, EDTA, is ignored because there is no profit potential for the drug companies compared to the new compounds that they can patent. Developed in Germany in 1932, EDTA has long since lost its patent protection for any pharmaceutical company. American drug patents last only seventeen years, and the new calcium blockers are the only anti-calcium agents you'll see promoted to the physicians and through them to the American public. Vested interests have largely kept chelation therapy from the American people. When they have trouble with clogged arteries themselves, some establishment physicians will take the treatment, but they rarely provide it for others for fear of stepping out of the medical mainstream. These doctors are known as "closet chelators."

Alfred Soffer, M.D., of Chicago, former Executive Director of the American College of Chest Physicians, wrote an article in 1975 that appeared in volume 233 of the *Journal of the American Medical Association*. It is sent routinely by the AMA to everyone contacting the organization for information regarding EDTA chelation therapy. This article, "Chelation Therapy for Arteriosclerosis," fails to discuss any of the research done that favors chelation therapy. It also provides mostly outdated information regarding the potential danger of the kidneys.

Most insurance companies today will not compensate holders of their health insurance policies when chelation treatment is taken for hardening of the arteries. The patients are denied any reimbursement. The insurance industry has labeled the therapy "not usual, reasonable or customary." It generally refuses payment on those grounds even if, because of chelation therapy, the patient avoids tens of thousands of dollars in bills for bypass surgery (which insurance does pay). Somehow cardiovascular surgeons have always seen to it that they get paid. And somehow heart and blood vessel surgery has never been required to go through the same rigorous and extensive testing that opponents want chelation therapy to undergo before it becomes "officially" recognized and thus payable by health insurance.

In 1910 Nobel laureate Alexis Carrel, M.D., indicated that human beings are genetically programmed to live 120 years. Russian scientists today have come to the same conclusions as Dr. Carrel. Yet we live a little more than half the number of our possible years, on the average, because hardening of the arteries from excess calcium brings us to premature death.

Johan Bjorksten, Ph.D., of Madison, Wisconsin, the world-renowned researcher on antiaging therapy, whom I mentioned previously, has identified the process of cross-linkages among the body's molecules as a major contributor to the aging process. In *Rejuvenation* (September 1980), Dr. Bjorksten wrote that chelation therapy has been shown experimentally to reverse cross-linkages due to the excessive accumulation of minerals such as calcium and aluminum. He suggested that chelation therapy offers the greatest promise of significant breakthroughs in life extension research.[3]

At the North State University of Texas in Denton, Texas, Foster Magill Walker, Ph.D., researched his 1980 doctoral thesis with EDTA. He definitely proved that EDTA chelation therapy lowers the level of calcium in the arteries of laboratory animals. The first public disclosure of Dr. Walker's work was offered during his presentation before the May 21–23, 1982 semi-annual scientific conference of the American Academy of Medical Preventics in Dallas. It was a confirmation of Dr. Bjorksten's declarations of life extension using the treatment, since Dr. Walker's laboratory animals, which were given EDTA, lived 25 percent longer than usual.

EDTA chelation therapy, mineral and vitamin supplementation, regular daily exercise, and other procedures recommended by the holistic physicians of the American Academy of Medical Preventics can be your instruments in achieving a longer and healthier life.

CHAPTER TWO

Health Problems Chelation Therapy Corrects

The chelation answer for enhancing your cardiovascular system comes about because ethylene diamine tetraacetic acid (EDTA) has the unique ability to form metal ion complexes. It removes these metals from the body. EDTA, upon being injected into the bloodstream, binds with the circulating unbound serum calcium. It also reaches into the smooth muscle cells that are part of an artery to form calcium-EDTA and other divalent metal- or mineral-EDTA complexes, such as lead-EDTA, cadmium-EDTA, and others. These actions happen at a blood pH of 7.35, which is most advantageous for the blood calcium to come out of solution and leave the body bound by EDTA.

By the time an adult living in the United States or in most of the developed countries of the world, where hardening of the arteries is running rampant, reaches the forties or fifties, there is a steady increase or accumulation in the calcium component in his or her blood vessels. Along with this extra calcium there is a marked decline in forty-six of the ninety-eight enzyme systems in a person's arteries that are active in one facet or another of arterial metabolism. This extra calcium inevitably must interfere with cellular metabolism in the arteries and in many ways helps cause hardening of the arteries. However, the EDTA chelate directs its activities mainly to the diffuse ionic calcium, which is the particular component that affects those enzyme systems. As a result of the law of mass action, calcium, being a very dominant factor in the blood, generally comes out in relatively large quantities during the process of chelation.

Thus, EDTA tends to remove some of the calcium complexes that are inhibiting the arterial enzyme systems. It helps correct the shifted enzyme balance that is producing insoluble calcium that piles up and contributes to the formation of atherosclerotic plaque in the blood vessel wall. Among the twenty-one different benefits from EDTA chelation therapy, three of the primary actions of EDTA are:

1) It reduces the excess diffuse ionic calcium accumulation, which helps correct the inhibition of the enzyme systems of the arterial wall.

2) It tends to stabilize the intracellular membrane of the cells of the arteries, and thus protects the biochemical integrity of the cells. (Medical scientists now believe that loss of the lysosomal membrane allows release of the powerful lytic enzymes that are inside the lysosome. The enzymes then digest and destroy the cell, leading to cellular death.)

3) It assists in maintaining the electrical charge of platelets in the bloodstream and thereby helps reduce platelet-leukocyte aggregations and clumping and helps prevent the formation of abnormal or unnecessary blood clots.

The end result of too much ionic calcium buildup is the production of an oxygen deficiency state (hypoxia) that triggers off a vicious biochemical cycle of events leading to cellular dysfunction, molecular alterations, a gradual buildup of atherosclerotic plaques, artery obstruction, arterial spasm, heart attack, and finally death.

Using EDTA chelation therapy as a preventive procedure, this diffuse ionic calcium buildup in the arteries may be avoided, or the accumulated deleterious intracellular calcification from years of improper diet may be partially reversed. A number of different health problems can thus be corrected by a series of twenty to fifty intravenous injections in a series with the man-made amino acid being discussed. The following situations describe how EDTA chelation therapy works to resolve an illness or disability when other medical or surgical measures have failed.

FOR REMOVING THE EFFECTS OF HEART ATTACK

At age sixty-six, Charles J. Fonas, a business executive of North Huntingdon, Pennsylvania, was sitting in his corporate office one Monday morning in January 1976 when he was hit in the chest with what felt like a freight train. The pain came fast, took his breath away, and left him just as quickly. Fonas shrugged it off as indigestion.

Pain returned the following Wednesday, but again the executive talked himself out of his worst fears. "Aw, they're just chest pains!" he said. "Nobody in my family ever suffered from a heart attack."

But the pain that struck a third time on Saturday was so severe, Charlie Fonas finally took himself over to the local hospital. He underwent the usual tests, including an EKG, an X-ray examination, and a full medical physical. The young resident physician concluded that Fonas really did have just chest pains—"muscle spasms," he called them—and prescribed pain pills.

Fonas took the prescription faithfully but still felt anxious about his recurrent chest discomfort. In truth, he lacked confidence in the diagnosis made by the hospital resident. The next week Fonas visited his regular physician. Another examination and a second EKG disclosed that he did have irregularities of the heart. His physician hospitalized Fonas for three weeks, and the pain did not come on again.

Just eight months later, Fonas swung for a golf drive on the seventeenth hole, and simultaneously with the swing he was struck in the chest by a severe heart constriction. It felt as if he had hit himself with the golf club. He tried to ignore the pain and finished the game. After that, though, walking even one city block regularly brought on the pain of angina pectoris. Fonas felt a kind of severe aching burn inside his chest from any exertion.

"Let's have an arteriogram," his physician suggested. "The cardiologist will put a catheter up through your arteries and determine if you can have a coronary artery bypass operation performed to relieve the angina."

But the arteriogram report was practically a forecast of doom. The cardiologist explained, "You have a rather severe diffused arteriosclerotic process which makes you an exceedingly unsuitable candidate for coronary bypass. Blockages in your coronary arteries amount to 80 percent stenosis [a severe arterial narrowing] in certain areas, 90 percent in other areas, and 60 percent in a third area. Surgery is out of the question. About the only thing we can do for you is give you drugs: a coronary dilator and Coumadin, which is a blood thinner. That's all that medicine has to offer. You'll just have to take it easy, Charlie, and hope for the best!"

For his family physician, the cardiologist wrote a report that read in part: "The patient's prognosis overall does not appear encouraging. His EKG manifested an abnormal central infarction with T-waves indicating anterio-lateral ischemia."

Interpreted, these various statements meant that Fonas had anginal heart pain that came on from any physical effort because of the death of a portion of heart muscle and narrowing of the coronary arteries. The area of destroyed heart muscle and insufficient blood supply meant that for the rest of his life, probably a short one, Charles J. Fonas would be an invalid.

Like Mr. Fonas, middle-aged American men and women frequently are affected by the obstructing deposits of atheroscelerosis. Plaque deposits in varying amounts line the interior wall of arteries and veins, and are irregularly distributed in large and medium-sized vessels. As time progresses, severe plaque formation leads to reduction of the central arterial blood channel and can cause what has been called the "all-American" disorder—heart disease.

Charles J. Fonas was told that nothing could be done and was figuratively sentenced to a death watch until another heart attack would end it all for him. The man worried but he refused to sit in anticipation of the blow that would strike him down forever. He searched for and grasped at any opportunity to reverse his atherosclerosis when the opportunity presented itself.

Undergoing terrible anxiety and distress, Fonas looked everywhere and was willing to do anything to improve his condition. Even while taking the prescribed vasodilator and blood-thinning drugs, he found himself unable merely to walk across a room without feeling the sensation of choking chest pain. The deep, aching burn had gotten worse. Angina pectoris made him a cripple unable to do anything for himself. He was vegetating with physical inactivity. The man was getting prepared in all ways—psychologically, legally, financially—to die. He was saying farewell to friends, relatives, business acquaintances, and neighbors.

Then, from a friend, Fonas heard of EDTA chelation, this unrecognized, unheralded form of therapy given by only about three-tenths of one percent of all the practicing physicians in the United States. Chelation might be the answer—the only answer—to saving his life. To learn more, Fonas did a lot of long-distance telephoning around the country until he found a doctor who gave the treatment. Then the patient set up a series of appointments for a month's stay in the doctor's city, a thousand miles from his home. On the last day of October 1976, Fonas flew south to find a way to reverse the blockage in his arteries. The salvation for him turned out to be this chelation treatment.

"I went on the intravenous bottle," he told me, "and the first day I tried to walk I actually made a hundred yards. Then I stopped and rested for five minutes and then made another hundred yards. The following day after treatment I walked about six hundred yards with many stops in between. The third day I walked about a mile; the fourth, about two miles; the fifth, three miles; and each of these days I was out about two and a half hours, which means that I made quite a few stops to rest between walking attempts. I stopped at any sign of pain."

Fonas measured his progress daily by adding distance to his outing. Within two weeks he had no pain at all. He managed to go a twelve-mile distance in his two-and-a-half-hour time allotment for exercise, with a half-hour stop in between. To walk six miles an hour (a very rapid pace) in that way is exertion even for a normally healthy man.

"When I told the chelation doctor of my progress even he was surprised at the very fast recovery. He checked me over to make sure that I had no damage; and there wasn't any," said Fonas. "Now I walk as much as I want!

"I recommend this chelation treatment for anybody who has had a heart attack and is considering open heart surgery or who has poor blood circulation of any kind," he said. "It took me thirty treatments to get in the good shape I'm in now. I have been working up to fourteen hours a day. In 1977, I helped to move my manufacturing plant from Monroe Hill, Pennsylvania, to Latrobe, Pennsylvania, and that was a tremendously strenuous job, but I felt great!"

Charles J. Fonas reaffirmed how he felt when I checked with him again following our 1978 interviews.

TO BRING BACK THE ELDERLY FROM SENILITY

Charles H. Farr, M.D., Ph.D., of Moore, Oklahoma, formerly a medical toxicologist, is now in private practice in rheumatology as a subspecialty. He was exposed to the pharmacology of chelation therapy in 1958 while studying toxicology in relation to heavy metal poisonings. He read of the treatment in connection with its use for prevention of hardening of the arteries and was impressed with the results.

Upon entering clinical practice after a career in basic research in the evaluation of drugs, Dr. Farr was challenged by many elderly arthritics who suffered with hardening of the arteries of the brain. He decided to try the intravenous chelation method to treat this senile brain condition. After all, nothing else in modern medical therapy seems to work for altering the course of senility.

"The individual with obvious cerebral arteriosclerosis is confused, disoriented, doesn't know his name, and doesn't know what he is doing," Dr. Farr said. "We have no way of determining how much actual brain tissue death has occurred. It may be just a transient hypoxia [not getting enough oxygen to the brain] from the arteriosclerosis.

"Or it may mean that because of the lack of oxygen, the brain is completely dead in an area. In that event, no matter how much oxygen the brain finally receives, it will do no good." Dr. Farr said.

"I treated an eighty-year-old lady, Mrs. B.W., whom most physicians would have given up on," said the physician. "I visited with her on June 5, 1974. She had been admitted to the hospital previously on several occasions for stays as long as thirty days. Each time she was in a state of total infantile withdrawal. She had cerebral arteriosclerosis.

"Nonconservative methods of all types that I had tried did no good for Mrs. B.W. She simply deteriorated further," he explained. "The woman was not cognizant of where she was, who she was, or anything else about herself. She usually would lie in bed curled into a fetal position, babbling like a baby, and defecating on herself.

"Although I knew my chances were poor in helping this lady, and her family concurred, we decided that chelation therapy was worth a try," he said.

"With just five weeks of the treatment by July 11, 1974, there was marked improvement in Mrs. B.W. During this period she became more oriented, felt considerably better physically, and she looked better. She knew what she was doing and where she was.

"One month later, in mid-August, my patient's son died in California. She booked passage on an airplane herself and flew to his funeral by herself. She returned completely unattended—a remarkable feat considering what state Mrs. B.W. had been in just two months before," said Dr. Farr.

The National Institutes of Health reports that senility is increasing among our older population. Approximately 15 percent of all people sixty-five to seventy-five years old and 25 percent of those seventy-five and older are presently senile, a total of about four million Americans.

This is possibly because each day, as part of the aging process, it is believed by many scientists, 100,000 brain cells die, the number actually being determined by how much pollution you are exposed to. The average person begins with about twelve billion brain cells, so that you would have to live more than 328 years before you lost all of them. But suppose you lost more than the usual allotment as a result of continuous pollution exposure, such as being the toll-taker on a bridge or always eating processed convenience foods; you would multiply your brain cell loss markedly. Speed up the neuron loss by five times, and by age sixty-five you could be totally senile, with insufficient brain cells left to recover. In fact, this is already happening to one-fourth of the U.S. population. People are suffering with brain cell deterioration as a result of impaired brain circulation from hardening of the arteries to the brain and direct brain cell damage by toxins.

The arteries give rise to enzyme systems that carry on their work and respond to arterial irritants or stimuli. These enzyme systems are quite efficient in youth, but with the advent of years, starting in the fourth decade of life, they begin to diminish in supply to about one-half of their previous activity. This reduction in enzymes decreases the repair processes of arteries so that they can't maintain themselves. It allows the influx of cholesterol to form atherosclerotic plaque. There is then an increased potential for platelets to become injured by hitting the roughened plaques on arteries, so that platelets become activated and "sticky" as they expect to find a wound which they are supposed to help heal.

These sticking or adhering platelets constitute the nucleus around which a clot may form. They can release a powerful substance that causes spasm and also promotes further smooth muscle proliferation. The sticking platelets may be the site then where atherosclerotic plaque begins accumulating sufficiently to diminish blood flow and obstruct the artery. The narrowed area of the central channel may then be the place where dangerous constriction or spasm of the artery takes place, leading to angina pectoris or even heart attack.

Spasmodic constriction of the muscular layer of an artery can occur anywhere in the body, including in the arteries of the circle of Willis of the brain or the carotid arteries in the neck which lead to the brain. Medical scientists and gerontologists consider it most likely that spasm hits at an obstructed area where plaque has formed.[1]

A suspicion of arterial spasm, incidentally, explains why an individual having a heart attack or stroke really has no greater narrowing or obstruction by plaque of the involved blood vessel today than he did yesterday. In many cases, it is simply superimposed blood vessel spasm happening in the narrowed area producing a relatively complete starvation of the part of the body fed by that blood vessel. The result could be a stroke if the blood vessel serves the brain or a heart attack if the blood vessel serves the heart, unless some calcium-blocking treatment such as chelation therapy is given—which can stop the spasm.

Cerebral blood flow studies using radioisotopes indicate that when this relative starvation is continued for a period of months there will be lasting impairment of thought processes such as short-term or recent memory and judgment. These brain dysfunctions are some of the symptoms of senility.

Amazingly, chelation therapy may return blood flow to these areas of the brain even though semi-starvation of its cells has gone on for years. There can be an associated restoration of memory in many cases of senility that have long since been considered hopeless.

REDUCING THE INCIDENCE OF PICK'S ATROPHY AND ALZHEIMER'S DISEASE

There is a special form of senility that develops among forty- or fifty-year-old individuals. This premature senility is called *presenile dementia,* and it may come from Pick's atrophy, a circumscribed or local shrinking of the brain cortex, which produces mental deterioration, loss of memory, physical inability to coordinate, and other effects that can accompany profound senility.

Alzheimer's disease is another form of premature senility that affects individuals in their fifties or sixties or even earlier. It is accompanied by senile-like symptoms of a psychotic forgetfulness, personality change, disorientation, and confusion. It differs from Pick's atrophy only in that Alzheimer's disease causes shrinkage throughout the brain, with degeneration of the nerve cell fibrils and senile plaques. Alzheimer's disease hits a person at an older age. It is associated with *Alzheimer's sclerosis.*

Most of the time, we don't think of younger people being senile. It is old Mrs. X or elderly Mr. Y who has symptoms of senility. Americans have grown to accept and even expect senility in the aged. When we see someone alert, active, and productive in his nineties we look upon that person with admiration and point him out as a marvel. The fact that we watch in wonderment when an older person continues to conduct himself creatively is testimony that someone not suffering from hardening of the arteries is the exception. This viewpoint is illustrative of acceptance in our Western culture that hardening of the arteries is a usual and customary condition for the twilight of our lives.

Chelation therapy is known not only to bring back the elderly from senility but also to reduce the severity of symptoms seen with conditions like Pick's atrophy and Alzheimer's disease. Some of these conditions are now found to be due to arteriosclerosis and some are due to excess aluminum in the brain. Chelation therapy has the ability to help in both of these conditions. There are cases where unfortunately brain cells have actually been destroyed and replaced by scar tissue, as after an old stroke. Chelation cannot help such problems of brain scarring, but the treatment may help prevent further difficulties, such as another stroke.

It appears overall that at least 50 percent of elderly people with senility problems show improvement after chelation therapy. Of course, all the necessary medical tests must be done to be sure the pathology is not brain tumor, chronic drug overdose, or other causes unrelated to poor brain circulation or excess heavy metal accumulation in the brain. These are the specific conditions resulting in senility in which chelation therapy seems to offer the greatest potential benefit.

A forty-six-year-old building inspector had been admitted by his physician, S.T. Talbott, M.D., to Doctors' Hospital, Coral Gables, Florida, for study of degenerative hardening of the arteries. The patient, whose initials are W.F., suffered from severely progressive weakness and cramping pains of his legs. His problem had developed over several years and now resulted in a total inability to walk.

The arteries in his head were quickly degenerating also. The building inspector experienced frequent loss of memory, could not form ideas, and was unable to make even the simplest calculations needed in his work. During what should have been his strongest and most productive years, the middle-aged man was growing prematurely senile. He seemed to be progressing into Pick's atrophy, arising out of his being affected somewhat by Alzheimer's disease.

W.F.'s blood pressure had increased to a dangerously high 230/130 (normal may be 130/80 for a male that age). He had no detectable pulses in his legs, and his toes were cramped into constant flexion, which further hampered his walking.

A neurologist called in by Dr. Talbott diagnosed extensive occlusive arterial disease (hardening of the arteries) with probable infarction (local death) of brain tissue.

A sophisticated diagnostic device used by another consultant revealed that W.F. also had severely narrowed and plaque-filled arteries in the trunk and legs. Plaque is usually seen in an artery as a fairly well demarcated yellowish collection of material on its inner wall surface comprised of various blood components including deposits of blood fat and other metabolic products. All these components often are held together with a matrix of calcium, scar or fibrous tissue, and platelets. In moderation, this may be considered "aging" by some; if accelerated, it represents the arterial plaque of arteriosclerosis or atherosclerosis. Such plaque caused a blockage, leading to reduced blood flow in the main branches and small branches of the patient's brain arteries and the superficial arteries of both his thighs as well.

Furthermore, the blood platelets traveling in this unfortunate man's bloodstream would hit this plaque. Thus injured or sensitized, the platelets were made more susceptible to releasing their powerful vasoconstrictive and clotting substances (prostaglandins such as thromboxane A_2). These vasoconstrictive substances induced an even greater spasm with additional obstruction of the blood vessels in already critically narrowed areas. This cycle of pathology led finally to essentially near starvation of his brain.

Dr. Talbott, with the concurrence of the various medical consultants, gave the man medicine to bring down his exceedingly elevated blood pressure, but, as happens all too frequently, the drugs brought on severe side effects. W.F. went into violent body tremors, so that he completely lost his coordination. After a month of these harmful side effects from his medication, the patient was seriously deteriorated and practically dead. A vascular surgeon thought he was too far gone for surgery.

Dr. Talbott was at the end of what he could do for the man by orthodox medical methods. Then he hit on the chelation treatment that relatively few doctors had heard anything about and only a handful have used. He called in another consultant to administer this therapy.

After three treatments the patient was walking normally. All muscular spasm and pain in his feet and toes had disappeared. He began to remember things and made plans for the future. Within twelve days W.F. was so much improved that he was allowed to leave the hospital to spend a weekend with his wife and children. In twenty-six days the Alzheimer's disease was gone, and the man's brain function had returned completely along with strength, motility, and coordination of his limbs. He was discharged from the hospital altogether.

Having received thirty-six treatments, W.F. was sufficiently improved in body and mind that the Coral Gables Department of Health approved his return to full-time work. He has not missed a single day at his job since.[2]

THE ELIMINATION OF INTERMITTENT CLAUDICATION

At age seventy, John Hardesty, Sr., a tall, stooped man with a distended abdomen, suffered from marked swelling of the feet and legs, an enlarged heart and enlarged liver, leg pain during rest, and painful intermittent claudication.

Although Hardesty was a rather sick man, the intermittent claudication was what troubled him most. He could not walk more than twenty steps without being struck by severe leg cramps. They stopped him in his tracks.

Five years before, Hardesty had been the victim of myocardial infarction (death of part of the heart muscle), which had left him with a weakened heart. It had less cardiac muscle cells to do the work of pumping blood. Part of his heart was now only scar tissue. Therefore, he developed the common problem known as congestive heart failure.

The patient's medical treatment included a maximum dosage of digitalis, a low-sodium diet, and two or three weekly injections of diuretics to reduce the swelling of his body. He experienced persistent bouts of nausea, an aversion to food, and deep mental depression. This unlucky individual accepted that his suffering was only being prolonged, with no real hope that anything would ever change for him. Indeed, he waited just to die, denied even the joy of taking a walk, which Hardesty loved to do. He just could not because he was too weak.

Then the late Carlos P. Lamar, M.D., of Miami, Florida, began the patient on a self-help program. He also administered the controversial intravenous chelation treatment. Hardesty's appetite returned in a few days, and in a few months he regained the weight he had lost previously. His sense of well-being was restored as well as the arteries throughout his body, previously rigid and overburdened with plaque and calcium, began to soften and open up. This was indicated by increased urinary excretion of calcium and enhanced circulation of the blood. Best of all, he found himself taking long walks—slowly at first and then with longer, faster strides. His intermittent claudication had disappeared completely.

Hardesty threw away his digitalis and discontinued the diuretics. At the end of three months the man's cholesterol level was nearly normal, swelling of his abdomen was gone entirely, pain in the legs was totally gone, and his enlarged liver and heart had shrunk back to their usual size. He could now walk as long and as far as he wanted. Hardesty felt like his former vigorous self again.

FOR REVERSING DIABETIC GANGRENE

After enjoying an outing in very hot weather during the fall of 1975, Roland C. Hohnbaum, D.C., then fifty-four years old, a chiropractor in Richmond, California, noticed that an ulcer had developed on his left foot. Such an ulcer was highly dangerous for Dr. Hohnbaum, since he is a long-term diabetic. Even with knowledge of the ramifications of diabetic ulcers, the chiropractor told himself, "This can't happen to me. I'm different!" But he was not different, and the gangrenous course of arteriosclerosis complicated by diabetes began its insidious creep from his toes upward.

"It developed worse and worse, and finally I was forced down and was flat on my back for about two months," said Dr. Hohnbaum. "I had an internist at the Alta Bates Hospital in Berkeley, California, whom I had a lot of confidence in. He took care of my case in the beginning but then he became frightened, saying that I'd have to go to the hospital because I was going to lose my toes."

Quite frightened himself, Dr. Hohnbaum understood it was time to make some major decisions. He decided against amputation for gangrene, since he knew that a decision to operate just *started* at the toes. Gangrene is known to spread, slowly and steadily, and eventually may require amputation below the knee, above the knee, or even at mid-thigh. Besides, the combination of hardening of the arteries and diabetes was by now showing obvious effects in his other foot, too. It meant having both his feet cut off.

Because he refused hospitalization and surgery, Dr. Hohnbaum was denied further treatment, first by his internist and then by the consulting vascular surgeon. Afflicted as he was, the patient had to engage in his own self-help program at home by bringing in dietary factors and anything else he could use to advance his health. He prayed that his condition might begin to show some improvement.

"I finally found out about Dr. Tang with the chelation therapy," the chiropractor told me while tears formed in his eyes, "and this was a lifesaver for me. I don't think that I can say any more than that I—I have feet under me now."

When Dr. Hohnbaum visited Yiwen Y. Tang, M.D., Fellow of the American Board of Family Practice, then of San Francisco but now located in Reno, Nevada, he arrived on crutches and in pain. "The patient was in immediate need of amputation of his two legs below the knee," Dr. Tang said. "His life was in imminent danger. He could bear nothing on his feet—not hosiery—certainly not shoes." After careful diagnostic studies, Dr. Tang began Dr. Hohnbaum on EDTA chelation therapy.

"The pre-chelation tests also indicated clogging of my carotid artery, making me a prime target for stroke," said Hohnbaum.

Within a week of having the first treatment the patient could put on socks. By January 1976, two months after the last of fifteen chelations, he wore shoes and returned to work full time. The chiropractor is today completely healed, with no evidence of anything having happened to his feet. His diabetic gangrene was completely reversed.

"Post-chelation tests show my carotid artery now clear of occlusion. There is no doubt that the atherosclerosis which accompanied my diabetes has abated," he concluded.

Dr. Hohnbaum's case and a hundred others like his have changed medical thinking in this country about diabetic gangrene. The condition had previously been considered irreversible, but diabetic ulcers and gangrene now can be cured by chelation therapy.

Roland Hohnbaum added a postscript to his story. He said, "I am still haunted by the flat statement made to me by a vascular surgeon whom I had met after my feet were all healed. Upon seeing the photographs of my gangrenous feet, the surgeon said, 'If you had been my patient and your feet looked like these pictures, there is no question—I would have amputated!' "

I interviewed Dr. and Mrs. Hohnbaum in Dr. Tang's office a couple of years ago, as the patient was undergoing his annual series of five chelation treatments. Mrs. Hohnbaum told me they were on their way that evening to pursue their weekly pastime—ballroom dancing. "Roland loves to dance," Mrs. Hohnbaum said.

TO RESTORE IMPAIRED VISION

Reynolds Hall, age fifty-seven, of Melbourne, Florida, is blind in his left eye as a result of a childhood accident. While shaving on the morning of June 23, 1973, he suddenly went blind in his right eye as well. "When my right eye went out, I couldn't see the mirror in front of me," Hall recalled to *Today* staff writer Howard Wolinsky.[3] (*Today* is a Gannett newspaper published in Brevard County, Florida.)

Hall said the vision in his right eye improved somewhat on its own during the time he was visiting physicians who specialized in eye and brain surgery. He was still considered legally blind when he was admitted to Brevard Hospital in Melbourne.

"After the testing was done, the docs said there was nothing they could do for me," Hall explained.

Robert Rogers, M.D., of Melbourne, Hall's family physician, suggested chelation therapy, since Hall suffered from hardening of the arteries. His condition had been so severe that the circulation of blood to his left leg had clogged completely and the leg was amputated in 1970.

Dr. Rogers suggested that the combination of obstruction and vasospasm apparently was blocking arteries to his eye located deep in Hall's head, causing the blindness to his remaining eye, a condition known as *macular degeneration*. The physician gave the patient intravenous chelation therapy while he was in the hospital in hopes of removing some of the blockage.

"After I took seven treatments, the miracle took place. I could see," said Hall. "I was looking across the room and about fifteen feet away I could read some tiny writing on the TV. It said 'Hospital Communications Systems.' I got real excited and called Dr. Rogers. When the doctor got to the hospital *he* couldn't even read the words [the print was so small]."

When an ophthalmologist checked his vision, it was found that Hall was 20/15 in his right eye—better than 20/20. In a letter that he wrote, another physician, Robert Sarnowski, M.D., a neurosurgeon in the Melbourne area, attested to the "dramatic" improvement in Hall's vision after chelation therapy.

The reason? Hall's form of blindness, macular degeneration, appears to occur from impaired blood flow to a portion of the eye. Chelation therapy brings back the circulation to such an eye affected with macular degeneration and thus restores impaired vision even in eyes that are legally blind.

AS AN IMPROVEMENT FOR MEMORY

Warren M. Levin, M.D., F.A.A.F.P., of New York City, described his personal experience with giving himself chelation therapy. He said, "I went through a full preventive medical checkup, including cardiac stress testing, and was pronounced 'fit as a fiddle.' However, my father had died at the age of fifty-six with a coronary. Also, I'm short, relatively stocky, and had been eating what I now consider unhealthy food for the first forty years of my life—until I learned better. I know that I have been undergoing the same sort of degenerative changes of hardening of the arteries like everyone else. I think that in the past few years, by altering my lifestyle with regular daily exercise and better eating habits, I have slowed down the pathology. Nevertheless, I wanted to do more than just slow it down. I wanted to eliminate atherosclerosis from my body altogether, if I could. So despite my feeling in really fine health, I took a series of twenty chelation treatments in the spring of 1975.

"About eight weeks after I finished self-administered chelation, I suddenly realized that a very important memory change had taken place in me. My memory was much improved," said Dr. Levin. " I have always been blessed with a pretty good brain—high IQ and all that—and now I am able to remember long-unused addresses and telephone numbers. I had no problem with that kind of remembering. My annoyances would occur in the middle of a workday in this unbelievably hectic office—allergy tests being taken, chelations being given to patients, physical examinations going on, and other activities.

"I might be examining one person and have another waiting in my consultation room," Dr. Levin continued. "The telephones would be ringing. My nurse might ask me about service charges or appointments for patients—sixteen different inputs coming at me at once. Then I may walk into my consultation room from the examining room and have to stop to ask myself, 'What did I come in here for?' Often I'd have to turn around and go back to the examining room to see my patient or look at the chart, and then it would come to me. I'd realize, 'Oh yes, I was supposed to get that order blank or fill out the form on my desk.' "

Dr. Levin assured me, "That sort of memory lapse doesn't happen to me anymore. The whole pattern has modified dramatically. There is absolutely no question in my mind that my improved memory is a direct result of chelation therapy. I took a few more treatments in 1976, 1977, 1979, 1981, and more recently, and I will be taking chelation at fairly regular intervals indefinitely. It's the best *preventive medicine* that we have to offer in this country to eliminate hardening of the arteries for somebody in my state of health and at my age of forty-eight years."

OTHER BENEFITS OF EDTA CHELATION

What is this, some sort of miracle potion? A panacea? A treatment to improve the memory, restore vision, open blocked arterial channels, renew libido and sexual potency, reverse gangrene, relieve claudication pain, and rejuvenate the cardiovascular system in general—EDTA chelation treatment does all that? In a way! Chelation therapy appears to reduce pathology and improve physiology wherever the basic issue is an interference with the flow of blood to a cell, tissue, organ, or body part. In fact, any system of treatment that brings fresh blood and oxygen to the tissues can be expected to aid the healing process. Cleansing the blood by chelation does just that. When used properly, following the protocol of the American Academy of Medical Preventics, this treatment is extremely safe, scientific, and highly effective.

In their book *How to Survive the New Health Catastrophes,* H. Rudolph Alsleben, M.D., and Wilfrid E. Shute, M.D., explain some additional chelation therapeutic effects that occur with intravenous administration of the synthesized amino acid disodium EDTA.[4] This substance provides the following beneficial effects—it:

- Prevents the deposit of cholesterol in the liver
- Reduces blood cholesterol levels
- Causes high blood pressure to drop in 60 percent of the cases
- Reverses the toxic effects of digitalis excess
- Converts to normal 50 percent of cardiac arrhythmias
- Reduces or relaxes excessive heart contractions
- Increases intracellular potassium
- Reduces heart irritability
- Increases the removal of lead
- Removes calcium from atherosclerotic plaques
- Dissolves kidney stones
- Reduces serum iron
- Protects against iron poisoning and iron storage disease

- Reduces heart valve calcification
- Improves heart function
- Detoxifies several snake and spider venoms
- Reduces the dark pigmentation of varicose veins
- Heals calcified necrotic ulcers
- Reduces the disabling effects of intermittent claudication
- Improves vision in diabetic retinopathy
- Decreases macular degeneration
- Dissolves small cataracts

Moreover, William J. Mauer, D.O., of Zion, Illinois, a Diplomate in Chelation Therapy of the American Academy of Medical Preventics, has added to the list of benefits the patient derives from EDTA chelation therapy. From his own experience and the experience he has gathered from other chelating physicians, Dr. Mauer affirms to me that the treatment:

- Eliminates heavy metal toxicity
- Makes arterial walls more flexible
- Manages excess quantities of fat in the blood
- Prevents osteoarthritis
- Causes rheumatoid arthritis symptoms to disappear
- Has an antiaging effect
- Smooths skin wrinkles
- Offers psychological relief
- Assures the presence of adequate zinc in the blood
- Lowers insulin requirements for diabetics
- Dissolves large and small thrombi

Some major institutions have underwritten studies of the intravenous therapy, and scientific articles are published as a result of their underwriting. Grants-in-aid for research on EDTA chelation therapy have been given by:

1. The National Institutes of Health
2. The U.S. Public Health Service
3. The National Institute of Arthritis and Metabolic Diseases
4. The University of California, Los Alamos Scientific Laboratory
5. The American Cancer Society
6. The Charles S. Hayden Foundation
7. The Elsa U. Pardee Foundation
8. The Oscar Meyer Foundation
9. The U.S. Atomic Energy Commission
10. The John A. Hartford Foundation

11. The Providence Hospital Research Department, Detroit, Michigan
12. The Equitable Life Assurance Society, Bureau of Medical Research
13. The U.S. Public Health Department, Division of General Medical Sciences

The scientific literature search I have conducted indicates that there has been a high degree of success in achievement of therapeutic benefits through the use of chelation therapy.

A PERSONAL NOTE

At this point in my writing I must interject a personal note about what you have read so far and what is to come. Usually the conclusion that an objective observer draws, especially someone trained in science and medicine, like myself, is that more often that not anecdotal medicine should be looked at dubiously and validity established only from the published documentation, especially requiring double-blind and single-blind clinical studies. In the past I believed this to be true. But what must an investigative medical journalist do when he is exposed to anecdotes, patient stories, and one case history after another that reports potentially imminent death, blindness, amputation, paralysis, and other problems among people; and upon visiting those people to check their stories, he sees them presently free of all signs of their former health problems? This has happened to me!

About three hundred thousand individuals who were victims of hardening of the arteries are much changed. I have talked with about 285 of them. They have become *former* victims. Now these chelated people are vibrant, productive, youthful-looking, vigorous, full of a zest for life, and they enthusiastically endorse chelation therapy as the cause of their prolonged good health. Many thousands of them have joined together to form the seventy chapters of the Association for Cardiovascular Therapies, Inc. to bring chelation education to those not informed about the treatment.

I have checked most of what these chelated people claim and turned up not a single untruth. Whether the scientific method accepts anecdotal evidence or not, the information about the clinical effects of chelation therapy on hardening of the arteries deserves to be revealed. All persons whom I interviewed or corresponded with delivered their messages of recovery with ardor. They wanted to share their sense of renewed health and well-being with others who are suffering or will suffer. Consequently, I decided to bring a number of their messages to you in this book. You may want the same thing that these people have found—freedom from hardening of the arteries.

CHAPTER THREE

Why Arteries Harden and How EDTA Chelation Stops the Process

During the last half of the twentieth century, chelating physicians have practiced a form of medical alchemy. They've caused injurious calcium that was lodged in muscle layers of human arteries to convert itself for more beneficial purposes of strengthening bones and teeth. This misplaced calcium has moved from cells in soft tissue locations to the hard skeletal structures where it belongs. And any calcium excess that could not be utilized for some useful purpose was made to leave the body along with EDTA. Within the first day or two following the intravenous injections these chelating physicians have employed, the extra calcium and EDTA no longer are present to do any harm. They leave the body as waste products.

In fact, intravenous EDTA injection therapy restores a youthful flexibility to tired, old, rigid blood vessels. It rejuvenates the arteries and allows blood to circulate where clogging may have been present before. This chapter tells as simply as possible how chelation treatment works to soften hardened arteries. It is a highly complex and technical subject.

What is an artery, anyway? Is it a conduction tube that functions statically, like a plumbing pipe bringing fresh water to your kitchen tap? Not at all! An artery is actually a living, vibrant, thriving, pulsating, moving, reproducing, and reacting body tissue—an organ by itself. Additionally, each artery is a biochemical manufacturing plant for many of your enzyme systems.

As I mentioned in Chapter 2, ninety-eight different enzyme systems originating in the arterial walls create thousands of enzymes to counteract the tens of thousands of stressors we meet throughout our lives. The enzyme systems need coenzymes such as certain vitamins to carry forward their biochemical reactions. Enzymes also require minerals, nucleic acids, and other substances to function effectively. For example, an arteriole (the smallest size artery) requires vitamins C, A, E, D, and all the B-complex vitamins, as well as other nutrients. It must have a host of trace minerals such as selenium, copper, vanadium, and manganese (to name but a few), along with some minerals in larger quantities, including magnesium, sodium, potassium, iron, zinc, and many more. The arteriole definitely must have DNA and RNA (nucleic acids) to permit basic enzymes to operate. If any of these components are absent or are present in wrong amounts, our enzymes will no longer function properly and the arterioles will suffer damage.

By age forty-five or fifty-five (at least among people in Western countries), approximately forty-six of the enzymes manufactured in a healthy arterial wall go into decline. This is a major but little-known factor in hardening of the arteries. As the various enzymes decrease in viability, artery function tends to decrease, and tissues begin to be starved for oxygen and other nutrients. Any artery, small or large, that begins to diminish the biochemical interactions among its various components will consequently begin to lose its natural barrier against hardening of the arteries. Then internal formation of the characteristic lesion, the atherosclerotic plaque, a lumpy thickening of the interior arterial wall, begins to narrow the passageway and starts a long, complex process that eventually may lead to a blood clot or spasm that can ultimately completely stop blood flow through the artery.

The disability and death associated with deteriorated arteries is particularly ironic because a growing body of evidence indicates that the entire hardening and obstructive process is almost completely preventable and substantially reversible.[1] A number of published reports based on fundamental animal research autopsy observations, and on evaluations of the heart, brain, and peripheral circulation with sensitive diagnostic devices before and following improvements in diet, exercise, and administration of chelation therapy, have indicated remarkable improvement in circulation and probable regression of human atherosclerosis.

Atherosclerosis, the form of hardening of the arteries that is the most common cause of illness and death today, initially involves the inner layer of the arterial wall, or *intima*. The artery wall has three layers, and the other two are the muscular *media* and the outermost *adventitia*. A medium-sized muscular artery, such as a main coronary vessel, consists of a series of concentric tubes of differentiated cellular and extracellular components in these three layers.[2] The media is one place where harmful excessive calcium may accumulate in smooth muscle cells and begin to make the deadly changes that start to show as angina, leg cramps, impaired memory, and other symptoms.

THE CAUSE OF A SERIES OF INJURIOUS EVENTS

The plaques of atherosclerosis may become covered with blood clots that form themselves into thrombi (obstructions to blood flow). These progressive changes lead to increased slowing of blood flow through the arteries and frequently may close the blood vessels entirely.

If thrombi blockage or severe spasm involves arteries that feed the heart muscle, the sick person is hit with a heart attack.

If this entire clotting process blocks cerebral arteries, a stroke occurs.

If arteries of the eye are affected, the hypothesis is that diminished vision and even total blindness occur. This process is known as macular degeneration. Atherosclerosis can attack the interior wall of any artery in the body.[3]

The various basic and contributing causes of this entire series of injurious events that leads to impaired circulation are not yet completely understood, but a great deal of progress has been made recently and a significant contribution to our understanding was recently gained from the use of the electron microscope. In an article with the title "The Origin of Atherosclerosis," published in the February 1977 *Scientific American*, Earl P. Benditt, M.D., and his colleagues at the University of Washington School of Medicine proposed that the plaques were all a proliferation of a single mutating smooth-muscle cell from the media of the artery. This "monoclonal hypothesis" of Benditt's is the basis of the discussion that follows.

It has been well established that almost all children show the first signs of hardening of the arteries in blood vessel walls, the fatty yellow streak. This is a characteristic of both Western man and more primitive adults and children. However, more primitive populations do not usually proceed to develop the typical arteriosclerotic hardening of the arteries that culminates in the heart attacks and strokes from sudden, complete obstruction of their blood vessels.

Our recent understanding of the atherosclerotic process has opened the field to new approaches to treat and prevent this pathological process involving half of all deaths in the United States. The latest breakthrough we have found is that an aberration of the body's calcium metabolism is a major part of the problem.[4]

Calcium floats as ions in the blood, but it may bind itself in a very loose way with fibrous and elastic tissue in the intima wall of the blood vessels. This calcium binding tends to decrease elasticity and compliance of the artery. Ionic calcium can be released and returned into the blood serum when called upon by natural calcium mobilizers such as parathormone from the parathyroid gland. In the calcium metabolism cycle approximately 55 percent of blood calcium is bonded to protein and 45 percent floats free in the ionic form. Theoretically and practically, therefore, a significant portion of blood vessel stiffening and even the size of atheromatous plaque should be reversible. All that we need is the internal mechanism by which to safely mobilize calcium out of our arteries. To accomplish this, we can stimulate the parathyroid gland by lowering the ionic calcium with chelation therapy. Such a stimulation mobilizes excess calcification from the blood vessels and provides us with a major breakthrough in the treatment of cardiovascular disease.

Excess calcification of our tissues is a natural body defense to many irritants from our internal or external environment. We now are discovering that parathormone, the parathyroid hormone, can remobilize the excess calcium from pathologic locations, such as the excess calcium that has accumulated in the cells of your arteries in arteriosclerosis. It even accumulates in your joints as a result of disease, including many types of arthritis. In other words, parathormone removes excessive calcium present in abnormal locations and puts it back into the bloodstream to raise the blood calcium level back to normal. It also helps to reactivate bone cells to remineralize your bones by gradually transferring pathologic calcium off your arteries and back to your bones, where you need it.

Thus, many factors in our current industrialized Western lifestyle may contribute to calcium being deposited excessively where it should not be, such as in the soft tissues, including your arteries and tendons. Better it should stay in your teeth and bones, where it belongs. Many of these lifestyle factors were first identified by Hans Selye, M.D., Ph.D., D.Sc., in 1956 in his classic book, *Calciphylaxis*. Since then, additional factors have either developed or have been identified. Numbers of them will be described in Chapters 5 and 8. This calcium then interferes with the normal function of the tissue where it has accumulated and will eventually lead to disease, as well as contributing to the aging process. This spontaneous calcification goes on in our bodies constantly, but much more rapidly and extensively today because of our seriously imbalanced typical Western-type diet.

CALCIUM CREATES PATHOLOGY BY ITS SCLEROSING EFFECT

Unfortunately, with calcification taking place automatically anywhere in the body, the only time we will recognize it, (unless the doctor does a hair mineral test to look for it) is when symptoms begin to develop as a result of its sclerosing, or hardening effect. If calcification of nervous structures such as the motor tracts of the spinal cord occurs, the victim may develop or be diagnosed as having *amyotrophic lateral sclerosis*, or Lou Gehrig's disease. The patient with Lou Gehrig's disease will show symptoms that include progressive muscular atrophy, increased reflexes, fibrillary twitching, and spastic irritability of muscles.

If calcification of an inflammatory origin takes place in one's connective tissues,[6] the condition is called *progressive systemic sclerosis*, or scleroderma. Scleroderma is characterized by thickened collagenous fibrous tissue, with thickening of the skin and adhesion of underlying tissue, especially of the hands and face, but the skin thickening or hardening can occur anywhere over the body.

If calcification or sclerosis makes inroads as plaques in the brain and spinal cord, the patient may be diagnosed as having *multiple sclerosis* and suffer symptoms that show up chiefly in early adult life, with characteristic exacerbations and remissions. They include paralysis, tremor, nystagmus (rhythmical oscillation of the eyeballs), and disturbances of speech, although there are many other contributing causes to this all-too-common disease, including possible virus and even food sensitivity problems, but excessive calcium deposition in the nervous system may have helped set the stage for these problems.

Finally, if the calcification and sclerotic hardening occurs in the heart or blood vessels (primarily involving the arteries, although the capillaries and veins can be affected, too), doctors diagnose the individual as having developed atherosclerosis. Almost all of us in America get atherosclerosis by age twenty to forty today.

As indicated, excessive calcification of vital organs or tissues comprises a significant contribution to the underlying pathology of many of the degenerative diseases that we now see so frequently. Hardening of the arteries is the most common of these degenerative diseases. In fact, of the ten big killers among the people of the Western societies, hardening of the arteries may be a predisposing factor for six of them. In the order of incidence they are: heart disease, cancer, stroke, diabetes, cirrhosis, and arteriosclerosis. The other four, accident, pneumonia, suicide, and infant death, are generally unconnected to hardening of the arteries.

Hans Selye showed how blood calcium may be the precursor of the atherosclerotic plaque lesion. Excess levels of this mineral impair an artery's enzymatic ability to handle the fat material from the food we eat. Unfortunately, because of the tendency of American physicians to provide crisis care rather than preventive medicine, doctors usually wait to actually see symptoms develop (from the excess calcium in the artery) before they become concerned about dealing with it in a patient. By the time anyone has developed obvious vascular insufficiency symptoms, he is usually more than 60 to 80 percent obstructed. It is frequently too late then to begin simple preventive measures such as changing diet, instituting exercise, and other home care. More aggressive measures such as chelation or even surgery may be needed. Doctors will frequently notice this arterial calcification on your X-ray films. A calcium deposit must already be one-eighth of an inch thick before it can be seen on X-rays. Again, this is rather late to undertake preventive measures, and chelation therapy may be needed—or you can continue to ignore the calcification, as most doctors have done for years.

Medical scientists finally are realizing that most of us are increasingly developing excessive calcium levels in our soft tissues and this accumulation is now being recognized as a major contributor to both the aging process and degenerative disease. Now, preventive medicine doctors can diagnose it years before you become ill by simply using the hair mineral test, and then taking the necessary steps to prevent and reverse it.

While hardening of the arteries has been around for 3000 years (it's found in Egyptian mummies), heart attacks and strokes as epidemics are new problems of modern civilization. In 1969, Paul Dudley White, M.D., said: "When I graduated from medical school in 1911, I had never heard of coronary thrombosis, which is one of the chief threats to life in the United States and Canada today—an astounding development in one's own lifetime." In 1982, every second person who dies in North America is killed by coronary thrombosis, myocardial infarction, or some other form of heart disease.

Yet, you can have atherosclerotic plaque in your artery and still find its wall to be soft, nonrigid, compliant, and nonaged. Hardening has not set in. The plaque may be tumorlike—just a gob of material comprised of collagen, fibrin, mucopolysaccharides, cholesterol, tissue debris, and several other items. Cholesterol is not the main villain, as it has been oversimplistically represented. It is excessive calcium that helps make the artery stiff and unyielding. Without excessive calcium and its associated cross-linkages and muscle spasm, even a seriously obstructed artery could still be soft and pliable and thus provide adequate circulation.

The cross-linkages that calcium and other minerals (such as aluminum) help to cause, produce a hard, tubelike, inflexible blood vessel. Such a blood vessel seriously impairs circulation and also may stimulate blood platelets into a state of super-excitement. Stimulated platelets are all ready to try to protect you against bleeding to death from some "injury," which the hardened and roughened inner arterial lining simulates.

The result is that these now excessively fragile platelets may release a prostaglandin—thromboxane A_2, which is intended to stop your bleeding from a wound by producing spasm in the injured artery, thus clamping off the blood supply so you cease the bleeding. The thromboxane A_2 substance puts the muscular arterial coating into even worse spasm. Spasm in a major artery, especially if it is already partially obstructed by plaque, can completely cut off all circulation to the area the artery feeds. Such double spasming can be the final spasm, which may kill you.

HOW AND WHY YOUR ARTERIES HARDEN

As I mentioned before in describing Benditt's monoclonal hypothesis, three major developments take place in the formation of atherosclerotic plaque in a blood vessel wall: 1) Smooth muscle builds up to the point of resembling a benign tumor, 2) connective tissue lumps together in an excessive pile, and 3) intracellular and extracellular blood plasma components and debris accumulate at the site of plaque.

The first tissue change, smooth-muscle buildup, starts with a single smooth-muscle cell from the media. The cell spontaneously divides, and divides again abnormally to form a tumorlike growth that contributes a large portion to the bulk of the atherosclerotic plaque, particularly in its early stages. For many years experts thought this part of the plaque was simply "fibrous tissue" produced by the body in attempting to heal itself. Current concepts have changed.

The usually self-limited process of smooth-muscle growth tends to progress if either of two complicated circumstances occurs. There may be coexistent high cholesterol content in the blood (hypercholesterolemia) or something happens to damage the inner blood vessel lining. Either of these circumstances will cause thickening of the intima to an extraordinary degree, from 30 percent to 65 percent of ordinary thickness.

With hypercholesterolemia, the proliferating smooth-muscle cells become filled with fat droplets and extracellular fat (lipid) appears.

An injury to the intima may not lead to a lot of lipid accumulation, but it does provide an exuberant second wave of smooth-muscle proliferation. Abnormal proliferation is believed to be a response to mutagens that are present in our bloodstream, since mutagenic chemicals are now known to be routinely introduced into our water, food, and air. They include the chemicals we all read about that are thought to bring on cancer, such as substances in cigarette smoke, preservatives, pesticides, lead, and other things. None of us can escape exposure to them. But we can all become less susceptible to mutagens if we take certain precautionary steps, such as extra dietary fiber and vitamin C, which will be discussed at length in Chapter 7.

The next tissue change, which is connective-tissue lumping, begins and continues from the onset of plaque formation. The protein components of elastin, collagen, the mucopolysaccharides, and glycosaminoglycans accumulate in appreciable amounts. More connective tissue piles upon itself—a condition called *fibrosis*. Fibrosis is a reactive process the body uses to repair itself and probably is the most difficult aspect of plaque formation to reverse.

In the third phase of tissue changes, intracellular accumulation of blood plasma components, there is massive deposition of extracellular lipid. This lipid is the yellow-streaked, fat plaque aspect we often hear talked about in reference to atherosclerosis.

An enormous accumulation of cholesterol gathers in the smooth-muscle cell. The usual regulatory mechanisms operating in cells of the vessel wall are overwhelmed for a few reasons: (1) the normal enzymes are no longer being adequately produced; (2) excessive calcium in the cells is interfering with the normal metabolism; and (3) massive intracellular cholesterol-ester loading occurs. (Cholesterol, a fatlike material, and its esters, the bonding of alcohol and acid, are major components of plaque.) This is a necessary step in the development of the atherosclerotic plaque. The prior event to this "breakdown" of function is the marked increase in the permeability of the blood vessel wall. It leads to flooding of the underlying muscle cells with low-density lipoprotein (LDL) in a quantity many times greater than the quantity that usually passes through intact intima.

"Good" lipoprotein (high-density lipoprotein or HDL) is a fraction of relatively small molecular weight, rich in phospholipids. "Bad" LDL is a fraction of relatively large molecular weight, rich in cholesterol.

Harvey Wolinsky, M.D., Ph.D., and his colleagues at the Albert Einstein College of Medicine fed monkeys a 40 percent fat diet. This is a diet similar in content to that eaten by most Americans. After six to eight months of feeding, the monkeys had an average serum cholesterol of about 340 mg percent of blood, and mostly the bad LDL. Early atherosclerotic lesions were found diffusely in the animals' cardiovascular systems and focally in the coronary arteries. If the animals had not been sacrificed for autopsy, they would have died an early death from hardening of the arteries.

Writing in *Cardiovascular Medicine*, Dr. Wolinksy described other aspects of the formation of atherosclerotic plaques. He said that when the lining layer of the blood vessel wall has its integrity disrupted, this may be the event that sets the atherosclerotic process into motion—by the process of encrustation. In encrustation, the layer of elements forms a selective permeability barrier between the blood and the underlying vessel wall. It is a barrier dependent on intact cell-to-cell junctions that are quite impenetrable to most circulating proteins, and hopefully to most of the toxic chemicals we eat and drink and smoke today. This sort of "protection" is important so that certain blood vessel components don't come in contact with cellular elements in the blood—particularly platelets; otherwise abnormal clotting will occur.[6] Such clotting can produce a dangerous blood obstruction. The clot can completely obstruct circulation to a foot or brain, leading to death of that part (known as gangrene) or a stroke.

Sometimes our bodies are put under hemodynamic (blood movement) stresses in certain portions of the cardiovascular system, such as the blood vessels that supply the heart. When this stress takes place, there is an alteration of endothelial (inner artery lining) permeability. One such stress can be simply increased blood pressure, which can damage the inner lining by stretching the vessel wall and building up shear forces on it. The forces can denude a portion of the intima. Then our bodies send a carpet of platelets to cover the denuded spot within minutes. Platelet clumps can remain for many hours and even days—maybe permanently—and may become the basis for future pathology with the formation of typical arteriosclerotic plaques.

Additionally, when molecular chains of polyvalent metals, proteins, and nucleic acids combine together with other long protein molecules, an unnatural cross-link forms. The newly formed giant cross-linked protein no longer is able to function normally and cannot be split or hydrolyzed, as usually is done by enzymes present in the circulatory system.

The type of cross-linking I'm speaking of is illustrated by garden hoses that dry out and crack when they're old and no longer are soft and pliable, or by old rubber automobile windshield wipers that become hard and split. Cracked hoses and split windshield wipers have lost their elasticity as a result of a cross-linking, which is now recognized as one of the major contibutors to "aging." They are dried and brittle from the inability of molecules in the rubber to slide over each other.

The same sort of cross-linking-induced brittleness occurs in the protein chains of human beings, especially in the cardiovascular system. Consequently, along with the increasing calcium content that occurs in all of our arteries as we grow older, we find a loss of elasticity, reduced ability of the arterial wall to expand and contract, a tendency to become friable, and a loss of the artery's moisture content. These changes increase the *peripheral resistance* that your heart has to work against every time it tries to deliver another squirt of blood to your legs—thus making your heart work harder just to maintain normal circulation. Eventually, this increased work load is too much for the heart, and it begins to fail or you may begin to develop high blood pressure, so doctors may give you a drug such as digitalis or high blood pressure medication, which treats only the symptoms, not the causes. Doctors familiar with chelation would prefer to reduce the peripheral resistance your heart is working against, with the calcium-blocking and cross-linkage reversal effects achieved with chelation therapy. It deals with the actual causes of your heart failure or high blood pressure, such as spasm and aging cross-linkages that cause blood vessels to become too rigid, rather than just dealing with the symptoms.

Cross-linking is the basic way all human beings are. At a recent meeting of the National Institutes of Health it was decided that one of the accepted measurements of aging involves determining how elastic your tissues are. Biologically, if your tissues have give and can stretch without breaking, you are still youthful and not yet badly cross-linked, as we are when we become old. Dr. Johan Bjorksten, who conceived the cross-linkage theory of aging in 1941,[7] explained it in 1942 as follows:

> The aging of living organisms I believe is due to the occasional formation of tanning, of bridges between protein molecules, which cannot be broken by the cell enzymes. Such irreparable tanning may be caused by tanning agents foreign to the organism or formed by unusual biological side reactions, or it may be due to the formation of a tanning bridge in some particular position in the protein molecule. In either event, the result is the cumulative tanning of body proteins which we know as old age.[8]

Blood often contains a lot of cross-linking agents that reduce elasticity of blood vessels. They impart rigidity and hardness that eventually may lead to high blood pressure and/or cause little tears or breaks in the arterial walls, with subsequent leakage of blood into the surrounding cells. This leakage leads to plaque formation and is more likely to occur at points of maximal hydrostatic pressure, such as where twists or turns of the blood vessels are present.

AN "ABERRATION" OF CALCIUM FROM THE WAY WE EAT

Some of these dangerous cross-linking agents found in the human body help bring about the "aberration" of calcium metabolism that has recently been recognized as affecting coronary arteries and inducing blood vessel spasm.[4] Orthomolecular physicians and others using nutritional therapies point out that the modern but unnatural diet eaten by so many of us is largely responsible for the presence of these cross-linking agents in our bodies. The diet of Western industrial society, for instance, creates an imbalanced calcium-phosphorus ratio, from a normal healthy or optimal 1:1 to an abnormal 0.5:1.

Excess phosphorus comes from many sources in our unhealthy diet and has only recently been recognized as potentially dangerous and unhealthy. Most of us are consuming disproportionate amounts of calcium to phosphorus. For example, most carbonated soft drinks contain phosphates but do not have any balancing calcium to help us maintain the healthy 1:1 ratio. Nevertheless, the consumption of soft drinks in the industrialized countries is steadily increasing. We also get more phosphorus in our diet because it is a common food additive. Americans tend to eat large quantities of beef, which is forty-four times higher in phosphorus than it is in calcium.

All these factors contribute to the problem of a seriously imbalanced calcium-phosphorus ratio in today's diet. As a result, you are likely to suffer a significant loss of bone calcium, with an associated excess calcification of your soft tissues, especially in the heart and blood vessels, from ingesting high-phosphorus, low-calcium, and low-magnesium foods year after year. The paradox is that a reduced dietary calcium intake causes abnormal calcium deposition in your soft tissues, where it doesn't belong.

Jeffrey Bland, Ph.D., of the University of Puget Sound in Tacoma, Washington, has shown that taking a balancing calcium-magnesium supplement or eliminating the excess phosphorus-containing foods from your diet can provide a marked reduction in the problem. Avoiding phosphorus or supplementing with calcium and magnesium will help reduce soft-tissue calcium deposits, and allow you to have strong bones and teeth by keeping the calcium where it belongs. You won't be robbed of it each day from your bones and teeth while your body tries to accommodate to the excess phosphorus in your diet.

All preventive medicine physicians today should assess your average weekly diet with the assistance of a computer. Such an assessment shows what your calcium-phosphorus ratio is. The physician should also get a hair mineral test to be sure you are not in negative calcium balance, which shows up first as elevated calcium levels in your hair. Negative dietary balance can thus be detected before your teeth or bones are dangerously demineralized and weak—and long before your arteries or joints become obviously calcified. It will be noticed long before you start to develop symptoms, which doctors often treat by masking the underlying problem with pain-killing drugs such as aspirin.

Another problem that may help cause aberrant calcium to accumulate in Americans' blood vessels is our excessive average vitamin D intake. Nutritionists now find that we are routinely ingesting over 2000 units of vitamin D_2 in our food every day, which has been put there by food processors. In a recent lecture, Dr. Fred A. Kummerow of Woodside Research Laboratory, University of Illinois, reported that laboratory animal research he carried out indicated that this intake, which is the average for the United States, regularly induces animal blood vessel lesions. The lesions appear identical to those seen in the hardening of human arteries. We get these excessive quantities of vitamin D_2 from dairy products, meat, and the processed food industry's vitamin D "fortification" programs. Such high levels of vitamin D_2 are dangerous to our arteries, concluded Dr. Kummerow.

Another potential contributor to our "aberrant" calcium problem is the fat consumed in the so-called prudent diet, recommended by the American Heart Association. In the prudent diet 44 percent is from vegetable fats and about 66 percent is processed fat containing "trans" fatty acids. The belief among physicians using nutritional therapy is that trans fatty acids are potential causes of both hardening of the arteries and cancer. Trans fats are the types, present in margarine, that can bring about altered cell membranes and dysfunctioning enzymes. These alterations of tissues may allow intracellular calcium to accumulate to further aggravate the calcium metabolism aberration I am emphasizing. In addition, chemically changed trans fat definitely increases blood cholesterol levels. A 40 mg percent increase in serum cholesterol levels was reported in Romania when margarine was added to the population's diet to replace butter.

Eating an excess of refined carbohydrates, particularly sugar and starch, is also harmful to the cardiovascular system and may contribute to the "aberrant" calcium problem. Consuming refined foods may lead to an insufficient intake, as well as an increased need for essential nutrients such as vitamin B_1, chromium, and dietary fiber. A deficiency of these nutrients impairs carbohydrate metabolism, causing elevations of blood sugar and stimulating the production of excessive insulin in response to our high-sugar diet so that a prediabetic form of sugar intolerance commonly develops. This is sometimes simplistically diagnosed as "hypoglycemia." More correctly it has been called nutritionally induced chronic endocrinopathy by Dr. Jeffrey Bland, who has recognized and described the far broader aspects of this important and common problem.

Medical scientists at the Washington University School of Medicine in St. Louis recently found that high levels of insulin inhibit an enzyme in the cell wall responsible for helping to regulate proper intracellular calcium balance. Since the interstitial fluid outside the cell usually contains a thousand times higher concentration of calcium than is normally present inside the cell, this excess insulin response to our improper diet simply opens the calcium floodgates into the cell by inhibiting this membrane enzyme from working. Excessive calcium thus will enter the cells, impairing metabolism, producing cross-linkages and premature aging, and eventually producing excessive arterial spasm. The spasm may present itself years later as high blood pressure, or even angina, or other symptoms of impaired circulation such as loss of memory, cold feet and impotence.

As I mentioned, Americans are dangerously exposed to high quantities of lead, cadmium, mercury, and aluminum in our water, food, and air as a result of industrial pollution. Such heavy metal toxicity impairs enzyme function and also has mutagenic properties. Cancer and heart disease are both aggravated by these toxic heavy metals.

Aluminum especially is responsible for producing cross-linkages in the elastin and collagen of our tissues, and thus seriously reduces the elasticity of blood vessels and neurofibrils, where it lodges. It's one of the main causes of an Alzheimer's-like disease in which the patient has premature senility with serious and disabling loss of memory. Aluminum and other toxic or heavy metals may disturb proper blood pH to produce an overacid serum state so as to lead to even greater aberrant calcium accumulation in cells. These dramatic discoveries so vital to your health will be discussed in Chapter 8.

Dr. Bjorksten says that lactic acid which is produced during exercise, is an excellent natural chelating agent. It could help to avoid cross-linking of the elastin and collagen of our arteries. But many Americans and others in Western society lead sedentary lives. We deny ourselves the lactic acid we normally produce with sustained physical activity. Few of us are using this natural chelator—exercise—which would also help take excess calcium out of our arterial cells. (Natural chelators are discussed at length in Chapter 7.)

Adequate dietary fiber can lower serum cholesterol levels and decrease the amount of mutagens present. Mutagens have a potential to damage cellular membranes, and they actually stimulate proliferation of muscle cells in the arteries (the basic lesion of arteriosclerosis). But we eat an insufficient quantity of natural fiber. Adding bran to meals is just not enough. We also need the high methoxypectin found in whole fruits such as oranges and apples.

Eating highly refined carbohydrates, large amounts of proteins, too much phosphorus, and excess vitamin D causes you to easily suffer a magnesium deficiency. Magnesium is needed to metabolize these foods. A study at the Downstate University Medical Center of New York has found that the increased incidence of heart attacks in patients living in soft water areas is due to magnesium deficiency associated with this soft water. Hard water protects because there is some magnesium present. The mineral magnesium can displace calcium and function as a natural calcium antagonist. Yet our intake of this key mineral has dropped by approximately 50 percent over the last forty years, while our dietary pattern has increased our needs for magnesium. Dr. Mildred Seelig in her important book, *Magnesium Deficiency in the Pathogenesis of Disease*, published in 1980 by Plenum Press, has documented very convincingly that simply supplementing this critically deficient mineral in our diets could alone significantly decrease the epidemic of cardiovascular and renal and skeletal disease we suffer as a result of our unbalanced Western diet.

Most Americans consume less than 100 mcg per day of the antioxidant selenium. Selenium is another essential trace mineral, whose lack further helps place people at risk for heart attack. Studies in countries such as Yugoslavia, where Yugoslavs consume approximately twice as much selenium as Americans, reveal that they have approximately one-fourth the number of heart attacks as compared to the United States. In fact, knowing the selenium intake of any nation helps you predict quite accurately the incidence of both heart attacks and cancer for the people in that country.

Clearly, there are many other potentially important nutritional areas requiring additional research. Medical scientists don't fully comprehend all the reasons for the apparent aberration in calcium metabolism that has now finally been linked to impaired circulation and hardening of the arteries. Meanwhile, the astute physicians offering chelation therapy as a medical specialty in the United States inform their patients about these food factors. They try to motivate people to at least follow the February 1980 Dietary Guidelines of the U.S. Department of Agriculture. The U.S. guidelines include suggestions to decrease your dietary intake of fat, cholesterol, sugar, salt, alcohol, caffeine-containing foods, and many refined foods. To this list, chelation doctors add restrictions regarding potentially excess ingestion of vitamin D, phosphorus, and the trans form of fatty acids. As I shall discuss in Chapter 7, they also recommend that you increase your consumption of fiber and essential trace minerals, particularly magnesium, chromium, selenium, zinc, and calcium to balance the calcium-phosphorus ratio, and take therapeutic levels of antioxidant nutrients such as vitamin E and vitamin C.

All of these various nutrients assist the body in preventing or reversing hardening of the arteries, which is truly a multifactorial disease process. Chelation therapy helps to reverse the damage that results from the aberration in calcium metabolism that the risk factors I described helped to induce. Chelation therapy works well in this regard, but this doesn't mean you, as the patient, shouldn't do some things at home to help yourself get well. Everyone should be involved in helping to keep these problems from developing.

HOW EDTA CHELATION STOPS HARDENING OF THE ARTERIES

In Chapter 1, I summarized succinctly the action of EDTA chelation therapy, but the mechanism is a lot more complex than my short disclosure indicates. EDTA goes through a basic but vitally important chemical mechanism known as *chelation* in the blood to help remove excessive calcium from artery cells. The EDTA acts like the claw of a lobster grabbing onto some tasty morsel from the floor of the sea. In fact, the word *chelation* is derived from the Greek work *chele*, meaning *claw*.

The chelate substance (a ligand) offers a firm, pincerlike binding of minerals involving two or more links. This binding occurs when the mineral—usually a metal— accepts electrons from the atoms of the ligand. Calcium is a bivalent metal, meaning it has two available links or binding sites. Chelation incorporates the metal or mineral ion into the EDTA heterocyclic ring structure (a heterocyclic ring structure is a compound in a combination of more than one kind of atom.) Calcium or other metals are encircled or sequestered by the complex EDTA ring structure. Such metals lose their physiologic and toxic properties. The metal is essentially imprisoned or sequestered, and it can now be excreted from the body in this bound, inert form.

Hemoglobin is a common iron chelate and chlorophyll is a magnesium chelate. Without chelation, these important substances could not properly function in the body, or in the plant.

The ability of EDTA amino acid to grasp or bind with pathological ionic calcium deposits is phenomenal. Entirely calcified legs and even calcified kidneys have been saved with this unique compound. Furthermore, it is catalytic in its power to excrete or mobilize calcium far beyond the dose of the chelating medicine itself. Some of this reaction takes place because of the parathormone response to the induced lowering of blood calcium during the infusion of the amino acid chelating substance.

As indicated, upon being injected intravenously, the chelating agent binds with the circulating unbound serum calcium to form a calcium-EDTA complex. Much of this complex is excreted in the kidney and bypasses the normal renal conservation of calcium. In a medical journal article Garry F. Gordon, M.D., and Robert B. Vance, D.O., wrote:

> Approximately 80 percent is cleared through the kidney in the first six hours, and 95 percent in the first 24 hours. Some of the EDTA-calcium complex may dissociate, and calcium is dropped when a metal such as lead or chromium, for which EDTA has a higher affinity, becomes available.[9]

Drs. Gordon and Vance went on to explain that the loss of calcium through the kidney produces a transient lowering of serum calcium with a concomitant decrease in serum phosphorus, as well as an increase in serum magnesium. All of these chemical alterations with minerals in the serum are beneficial to the heart muscle's contraction by means of an action potential across the cell membranes.[10] Also, there is a possible holding in balance of electrons.[11] Altogether, a patient experiences a number of improvements in myocardial contraction and generalized advantageous heart rate and function from the EDTA chelation mechanism.[12,13,14]

The body's homeostatic mechanisms are trying to bring the serum calcium back up to normal levels because chelation can decrease the ionic calcium levels to approximately 50 percent of normal levels during the four-hour treatment. It makes this attempt partly through elevating parathormone levels by increasing output from the parathyroid gland. The initial calcium loss from the blood is replaced from the relatively easily available calcium stores of the body.[15,16,17] Much of the replacement is provided by metastatic (pathologic) calcium that is deposited in scattered remote tissues, including the blood vessel walls. This pathologic calcium leaves the arterial cells where it is doing damage and goes back into the bloodstream, where it does some good.

The normal human body is capable of replacing calcium into the blood at the rate of 50 mg per minute. Therefore, up to seven times the accepted dose of EDTA could be safely given in as little as one-tenth the normal administration time. In other words, a patient could take intravenous EDTA in fifteen minutes instead of in three to four hours, as it is administered now—without serious ill effect for most of us—unless our kidneys are diseased or we have high levels of lead or other toxic metals present in our body. Having such a contraindication, then rapid treatment would be ill advised. However, I believe the treatment has the arteries work more efficiently when it is given over a three- to four-hour interval. EDTA is obviously quite a safe therapeutic aid. I shall discuss safety of use and any possible toxicity of EDTA further in Chapter 6.

The increase in parathormone levels, if frequently repeated, leads to activation of osteoblastic bone activity. *Osteoblasts* are the body's bone-forming cells that build the osseous matrix of the skeleton. New bone can form during or following the chelation process. Indeed, it appears from close examination of X-rays of bones before and after chelation that extensive EDTA chelation does stimulate new bone formation.

Bone formation from intravenous EDTA use apparently takes place in the following manner: The continued osteoblastic activity, which research indicates goes on for at least three months after parathormone stimulation, continues to remove calcium from the blood serum. Soft-tissue pathologic calcium in plaques or arterial cells continues to diminish in order to meet the need caused by the increased bone uptake of calcium. The therapeutic cycle continues long after a series of chelation treatments has been completed, and patients continue to improve all this time. EDTA takes up serum calcium and disposes of it as waste; parathormone activates bone-forming cells; bones grow stronger and require more calcium for their buildup; more atherosclerotic plaques and media cells give off loosely bound pathologic calcium to satisfy the parathormone-activated bone cells' demand; the hardened arteries continue to soften steadily in the process. The beneficial effects of chelation therapy have always been observed to go on for many months following treatment. It was known that bone formation continues at an increased rate and keeps causing dissolution of metastatic calcium—but it was not until this process was recently identified that EDTA chelation specialists could truly explain why patients always continued to improve—even after the treatments were finished.

The late Carlos P. Lamar, M.D., F.I.C.A., one of the original pioneers of calcium chelation in atherosclerosis, had described the reduction of calcification in the aorta. There was also simultaneous recalcification of previously osteoporotic vertebral and femoral bones. *Osteoporosis* is a porous condition characterized by scanty, thin, and reduced skeletal tissue.[18,19,20] Dr. Lamar mentioned seeing improved joint function as arthritic joint calcium deposits are decreased. For that reason, symptoms of arthritis are frequently dissipated by the treatment. The deformity in a joint may remain, but mobility returns and pain goes away.

Another advantageous side effect of the chelation mechanism, the strengthening of bones and teeth, can be explained in this way: As reversal of hardening of the arteries takes place, bones and teeth get stronger. Metastatic calcium that may have been avoided being grasped by the chelation claw goes into reinforcing existing bone under the parathormone-induced osteoblastic effect. Hardened arteries get softer and softened bones and teeth should get harder following proper EDTA chelation therapy—where appropriate mineral supplementation with zinc, magnesium, and other minerals is being given, dietary calcium phosphorus ratio is balanced, and active exercise is undertaken.

In effect, parathormone is providing an adjunctive form of chelation to what the EDTA is accomplishing. Parathormone causes those bone cells, which haven't taken calcium into the bones since you were a child, to become activated. Once again, the osteoblasts enter into an active growth phase; they become avidly hungry for calcium to move it from the blood serum to bone. For at least three months after a series of chelation treatments, you walk around leaching calcium out of your arterial walls and into your bones.

CALCIUM BLOCKERS COMPARED TO EDTA

On May 10, 1981, at the American Academy of Medical Preventics' semiannual Scientific Conference, held in Atlanta, Richard S. Grubner, M.D., Clinical Professor of Medicine, the State University of New York College of Medicine, President of North Miami General Hospital, Diplomate, American Board of Internal Medicine and Sub Specialty Board in Cardiovascular Diseases, stated that EDTA is the original calcium blocker—the parent compound of the new class of calcium antagonists. The calcium antagonists are known under the trade names of Procardia (nifedipine), made by Pfizer, Inc.; Calan (verapamil), made by G. D. Searle and Company; Isoptin (verapamil), made by Knoll Pharmaceutical Company, now owned by Hoffman-LaRoche; Cardiem (diltiazem), made by Marion Laboratories; and Angex (lidollazine), made by Johnson & Johnson. Just as Dilantin was the first of the anticonvulsants, aspirin the first of the antifever medicines, sulfa drugs the first of the antibiotics, EDTA chelation therapy is the first of these newly celebrated calcium antagonists. Interestingly, nifedipine is a poly-amino-carobxylic acid and EDTA is also a poly-amino-carboxylic acid. At least one of the recently developed new class of calcium blockers is therefore chemically the same as the proven and extensively studied EDTA. Even though EDTA has been legally available to Americans for nearly thirty years, it has been kept from most American heart disease patients all this time.

Calcium entry blockers are likely to find a firm foothold in the treatment of heart and blood vessel diseases. Second-generation calcium blocking agents are already in development. The first generation of these agents have had serious toxicity problems. That's why, despite worldwide experience with these cardiovascular drugs, the U.S. Food and Drug Administration has been slow to approve their use in this country. For example, one comprehensive review has noted that in 262 studies involving 8344 patients given verapamil, adverse effects were seen in 9 percent; among these, 3.6 percent had cardiovascular reactions such as lowered blood pressure and facial flushing; 2.2 percent had central nervous system reactions such as headache and dizziness; and 2.2 percent sustained gastrointestinal reactions such as constipation and nausea. Side effects with nifedipine have been worse, occurring in 17.3 percent of 5008 patients and necessitating discontinuance in 4.7 percent of the patients. Diltiazem's side effects were observed in 3.9 percent of 7884 patients, but they were more varied, including not only those side effects already mentioned but also slowed heart rhythm, skin rashes, nausea, and abdominal discomfort.[21].

The complex and differing mechanisms of action of the calcium channel blocking agents are not yet entirely understood. The various effects of this large group of cardiovascular drugs on heart impulse and on coronary and blood vessel smooth muscle indicate that they may be useful in many clinical problems. Clinical trials of nifedipine, diltiazem, and verapamil are promising, albeit still theoretical. Robert F. Zelis, M.D., Professor of Medicine and Physiology and Chief of the Division of Cardiology, the Milton S. Hershey Medical Center, Pennsylvania State University College of Medicine, Hershey, confirms that until recently, "the conventional wisdom among cardiologists was, first, that coronary vasospasm [marked contraction and narrowing of a blood vessel segment] as a cause of angina was an interesting but unverified hypothesis and, second, that even if documented, the mechanism was probably rare."

This sort of erroneous thinking began to be altered starting in 1959, when a classic report described Prinzmetal's angina, a "variant" form of heart pain characterized by unprovoked ischemic chest discomfort and, on EKG, a transient elevation of the ST segment not associated with death of a portion of the heart muscle. In the last few years coronary vasospasm has been amply documented as a pathophysiologic entity, one that occurs in a large minority of patients with angina. Prinzmetal's angina is chest pain at rest that is more severe—lasting ten to fifteen minutes—than anginal pain occurring on exertion. Attacks take place during sleep, especially in the early morning hours and perhaps in association with the deepest (REM) sleep.

The spasm of the coronary arteries is the cause of Prinzmetal's angina, for the most part. Increased coronary blood vessel resistance, perhaps from sympathetic nerve activity, is responsible. The mechanism of pathology overall is highly complex, and my explanation here is overly simplified so as not to bring in too many extraneous medical terms and factors. It's sufficient for you to realize that all muscle contracts and relaxes according to increases and decreases of intracellular calcium. But unlike striated muscle that activates the skeleton, smooth-muscle cells in the blood vessel walls rely primarily on the fluxes of extracellular calcium that moves across the cell membrane causing the muscle to pass from contraction to relaxation. Smooth muscle may also contract through calcium traveling along channels to increase intracellular calcium and bring about vasoconstriction. Finally, calcium serves as a cofactor with the ubiquitous protein calmodulin to determine the extent of smooth-muscle contraction of arterial walls.[22]

The key therapeutic event in all of this is transmembrane calcium flux—calcium ions that cross back and forth through cell walls lining the coronary arteries and other arteries. Since vascular smooth muscle is so dependent on extracellular calcium, agents that inhibit transport of calcium into the cell would be able to relieve coronary vasospasm—and they do.

The calcium entry blocking agents—EDTA being the first and still the most effective—prevent the elevation of calcium ion concentration in vascular smooth-muscle cells. Verapamil and diltiazem accomplish this effectively in about ten minutes. Diltiazem also slowly and steadily inhibits intracellular calcium accumulation by stimulating the sodium-potassium pump in the body, which, in reducing intracellular sodium, may also potentiate passive cellular calcium loss via sodium-calcium interactions. Indirectly, intracellular calcium will be decreased in vascular smooth muscle this way. Diltiazem takes sixty minutes to achieve its calcium blocking effect. Diltiazem seems to be most selective for blood vessels of the heart, whereas nifedipine causes widespread vasodilation throughout the body. Verapamil is also a general vasodilator.[23]

Similar to the way intravenous EDTA reacts with other drugs, orally administered calcium channel blocking agents, judging from experiences reported by European physicians, also do not interact. Concomitant administration with nitrates (nitroglycerine), betablockers (Inderal), digoxin, furosemide, anticoagulants, and other agents has been uneventful.[24]

Besides Prinzmetal's angina, verapamil, diltiazem, and nifedipine have each been effective in chronic stable angina. They reduce both attack frequency and nitroglycerin consumption in the patient and also improve his exercise performance. The mechanism involved is uncertain.[25]

To repeat briefly what was unknown to physicians even five years ago, heightened transport of extracellular calcium across the cell membranes lining the blood vessels and heart sets the stage for vasoconstriction. Smooth muscle and/or cardiac muscle contraction takes place. Spasm may close down blood flow and consequent oxygen infusion to the cells in direct need of such supply. Pain results. If constriction lasts for too long a time, death of the involved tissues ensues. With too much of the vital tissues being destroyed, the person dies. The calcium inhibitors or calcium antagonists or calcium channel blockers or calcium entry blocking agents (all synonyms for the same thing) may prevent such a series of events from taking place.[26,27]

There are some vast differences between EDTA and the other newly developed calcium blockers.

1) The calcium blockers apparently don't do any chelating, so they cannot rid the body of any toxic metals. Thus, the more recently developed anticalcium drugs may have to be continually administered to produce the desired response. They are not simply taken in one series of twenty to fifty injections, as with EDTA, and then discontinued because you probably won't need them any longer.

2) Some of the newer calcium blockers are metabolized in the body, with greater potential for toxic results. EDTA is not metabolized at all but

completely leaves the body without producing the more serious side effects reported with the new drugs.

3) The calcium blockers can induce heart block. EDTA has not been found to do this, possibly because its therapeutic index is so large.

4) The calcium blockers have no thirty-year history of extensive and safe application in children and the elderly alike, as does EDTA. Furthermore, there has been widespread intravenous use of EDTA in blood transfusions for years as the anticoagulant of choice. The current extensive use of EDTA in the diet as a preservative has provided doctors with experience from which they can definitely conclude that it is a very safe substance.

5) The calcium blockers don't remove lead and other toxic heavy metals from the body, as does EDTA.

6) There has never been a medical journal article showing that the calcium blockers will decrease the incidence of heart attacks by 50 percent and cancer by 81 percent for an eight-year interval, as I shall describe in Chapter 8.

7) There is no research with the new calcium blockers showing remarkable life extension of up to 50 percent, such as was shown with rotifers that were exposed regularly to EDTA.[28]

8) There has been no basic laboratory research to show the potential of reversal of cross-linkages with the other calcium blockers, as has been shown with EDTA.

In a total heart medicine market of about $2 billion a year, with current annual angina drug sales totaling more than $200 million, drug makers of the newly introduced calcium blockers intend to cash in big. But why have the potential benefits of EDTA chelation therapy been generally denied to most Americans all this time—since 1952? Primarily because no pharmaceutical company was willing to spend the required amount of money—about $30 million—to get it approved by the FDA for use in cardiovascular disease. The EDTA patents had expired before the drug's true potential was recognized. It is now an orphan drug. And the National Institutes of Health (NIH), which could have corrected the situation, continued to incorrectly claim that calcium was unimportant in vascular disease.

The involved pharmaceutical companies have been spending about $100 million now to get the FDA to approve the four new calcium blockers. Even though these drugs have much greater potential side effects and much less potential for benefit than EDTA for the American public, the money is going for FDA approval simply because the new compounds allow seventeen years on patent life. They have parents sponsoring them and aren't orphans. The drug makers fully expect the public will remain completely unaware that a far more effective, safer calcium blocker has been legally available all along. Until publication of this book, very few (laypersons and traditional physicians alike) were aware of EDTA chelation therapy for hardening of the arteries.

Millions of people have suffered vascular disease, its associated disabilities, and even death simply because they did not know about the EDTA intravenous injection technique. Patients' arteries became hardened. EDTA chelation was unavailable to reverse the pathological process and add years of potentially active and useful life for some of our most productive citizens.

CHAPTER FOUR

Noninvasive Blood Flow Testing for Chelation Therapy

Broadly speaking, there are two types of tests used in cardiovascular therapies: invasive and noninvasive. We'll speak briefly of invasive tests before describing in detail the preferred noninvasive tests.

Angiography is a diagnostic test that invades the body's interior. It provides instructive comparisons with noninvasive techniques. Although this medical-surgical procedure is well accepted by the orthodox medical community, of every 200 people undergoing coronary angiography, at least one of them dies on the X-ray table during this invasive examination. "The procedure is hazardous," says *The Medical Letter*, and warns against its overuse as a diagnostic test.[1] You may not consider the statistical rate of death so very dramatic until you realize that you may become the particular two hundredth person to become a death statistic.

Angiography, also known as *arteriography*, involves passing a catheter into the heart under fluoroscopic control and injecting radiopaque contrast material through the catheter into the coronary arteries. X-ray films of the dye flowing through the coronary arteries supposedly demonstrate narrowing or complete obstruction.

One of the problems with angiography, however, is that about half the patients are allergic to the dye. It has the potential of causing vasospasms or anaphylactic reactions and killing people, which definitely does happen. The average death rate for angiography, varying from 0.1 percent to 1.0 percent, depends on the doctor's experience, the age and health of the patient, and whether the death statistic takes into account when the individual died—at the time of the procedure or six months later from an embolism sustained when the angiogram was performed. Patients are quite likely to suffer from complications of the catheterization such as myocardial infarction, stroke, cerebral emboli, arrhythmias, thrombosis, arterial tear, and dissection at the site of arterial puncture.[2]

Be warned, then, angiography is not a test for you to undergo without weighing its consequences. Many experience terrible pain with the procedure. If you are sensitive to pain ordinarily, you may feel a very uncomfortable, hot, burning sensation while going through the examination, as if you were actually having a heart attack. The reason is that for four minutes during the test the blood is cut off to your heart—the same effect as in heart attack. There is only dye in the heart at this time and no blood. Some patients find the pain so unacceptable that they never would undertake an angiogram again under any circumstances. They describe their sensation as being suffocated almost to death.

Additionally, the patient is overirradiated. The amount of radiation to which you are subjecting yourself with this one invasive diagnostic technique is equivalent to having 250 to 300 chest X-ray examinations.

Then there's the cost! An angiologist charges anywhere from a low of $800 up to $1200, and that's just the doctor's fee. The hospital piles on its own charge of $3200 for an average minimum hospital stay of three to four days so that you can recuperate from this four- to five-hour procedure in which your life was in danger.

Perhaps worse than the personal discomfort, risk of death or complications, excessive radiation, and outrageous cost is that angiography is quite inaccurate. New studies reveal that it doesn't have much value for determining the state of your coronary arteries or other arteries. "The trouble is, there are no known standards," says Dr. Arthur Selzek.[3] The extent of your blockage is frequently guessed at by the angiologist reading the X-ray film.

According to William Stewart, Ph.D., of Rockville Centre, New York, a technical expert in nuclear medicine and Associate Professor on the staff of Lenox Hill Hospital in New York City, the angiologist is reading a two-dimensional picture of a three-dimensional artery. Dr. Stewart said, "The angiologist can't tell the true amount of lumen [blood channel] left open in the occluded artery, because he is actually looking at a narrowed coronary artery overall. He's unable to tell where the lumen ends and the intima begins. There is absolutely no way to determine the lumen location, for the radiologist measures plaque along with the lumen. What appears as a 50 percent stenosis [narrowing] in reality may be a 90 percent stenosis. This has been documented in studies performed at the National Institutes of Health.

"In addition," Dr. Stewart said, "with angiography each patient isn't being stressed as he should be, but rather his heart is being looked at while it's resting. As a result, in no way does the angiologist get a clear diagnostic picture of what is happening in the patient's heart at high velocity rates, which is when a heart attack usually occurs, because the angiogram is taken at rest."

Other cardiologists reading angiograms attest to the fact that the X-ray films will be even more difficult to decipher when the patient is obese. Moreover, experienced cardiologists and radiologists frequently differ in their interpretations.

From a double-blind evaluation of the reading accuracy of angiologists conducted at the National Institutes of Health, cited by Dr. Stewart, it was learned that the individual doctor reading the angiogram contradicts himself approximately 40 percent of the time. In October 1979, six radiologists located in the Washington, D.C.-Bethesda, Maryland, area, prestigious in their medical specialty of reading angiograms, were asked at different times and on different days of the week to diagnose the identical heart and coronary artery angiograms they had read before. The patient's name was blanked out and could not be seen. The mistakes these angiologists made relative to prior diagnoses ran 38 percent for underreadings and 42 percent for overreadings. The Harvard Group of Massachusetts General Hospital announced these glaring errors at the International Congress of Cardiology in New York City, sponsored by the American Heart Institute, in October 1980. In shock, most primary-care physicians present vowed they would not depend on angiograms as the sole diagnostic criterion.

In other words, suppose you subject yourself to angiography and are told by the cardiologist that you have an 80 percent closure of one or more of your coronary arteries and open heart surgery is mandatory. Because of the inaccuracy of even the most renowned angiologist in interpreting the angiogram, you have two out of five chances that the reading is a mistake and the report to your cardiologist is wrong. What the angiogram radiologist diagnoses on Tuesday, he will likely read differently on Thursday. Thus, if we rely only on repeat angiography after treatment such as chelation or surgery, no one actually knows if the state of your health is better or worse.[5]

So why should anyone ever submit to an angiogram? The conclusion must be that you undergo angiography with the disproportionate ratio of too much risk and too little benefit. While this catheterization technique has been somewhat beneficial for diagnosis in the hands of a highly skilled physician-technologist, it generally has an overburden of hazards when the technician is not so skilled. The application of angiography must immediately be limited markedly. The large number of catheter-trained doctors should go about applying their skills to noninvasive testing.

Inasmuch as vascular specialists did not have anything better for diagnosis until recently, angiography has been used to determine whether coronary artery surgery was indicated, especially in those people who suffered with severe disabling angina pectoris unresponsive to medical therapy. It may have been applicable also as a diagnostic determinant for preoperative evaluation for valvular surgery, and only *occasionally* for patients with recent myocardial infarction. A primary rule, often violated, in the use of angiography is that the advantage from its use must outweigh its perils.[6] Now we see from the NIH study that strict adherence to this rule would cancel most coronary arteriograms, since it is too inaccurate to justify its peril—unless you must have surgery.

Clinicians incorrectly used angiography to diagnose or exclude coronary artery disease in patients unlikely to be considered for surgery, while others considered the procedure too risky for such "routine" diagnostic use.[1] Most physicians, however, have employed the procedure for patients only when surgery was contemplated. It is relatively hazardous, but you don't know which arteries in the heart to bypass surgically without this study.

Depending upon arteriography, or angiography, as the sole criterion for determining the need for surgical correction of coronary artery lesions, which has been a common practice, is absolutely wrong. Perfect human specimens at age forty who run five miles every morning and can conquer treadmill tests of 185 heartbeats a minute will still frequently show documented abnormal angiograms (as much as 50 percent of healthy volunteers in a study), with at least 50 percent narrowing of the arterial lumen by age forty. Even so, the last thing the patient may need is life-threatening heart surgery. Chelation therapy would be ideal for these cases.

The American Academy of Medical Preventics (AAMP) maintains that its members will avoid prescribing angiography but instead will use noninvasive diagnostic heart and blood vessel studies almost exclusively. Because of angiography's risk to the patient's life, radioactive overexposure, unnecessary pain, excessive financial cost, and the inaccuracy of angiogram readings, AAMP physicians declare this invasive test may be employed in only the rarest of cases: where the patient has experienced a total failure of chelation therapy and potential coronary artery bypass surgery is the only thing left.

John Kirklin, M.D., a heart surgeon at the Mayo Clinic, said, "Surgery is always second best. If you can do something else, it's better. Surgery is limited, it is operating on someone who has no place else to go." Furthermore, heart surgeon Dr. Michael DeBakey stated on October 4, 1981, that he would be happy to see cardiovascular surgery "become unnecessary . . . I think there is now an increasing recognition for the thrust to be, from a clinical standpoint, finding ways of preventing disease. In other words, you try and help the population to learn how to maintain their health. There's a lot people can do. . . . The medical people have a responsibility to focus attention upon it."

HOW NONINVASIVE EXAMINATION COMPARES TO INVASIVE TESTING

Noninvasive cardiovascular diagnostic techniques are not hazardous like angiography and other invasions of the body's interior. Indeed, the benign noninvasive procedures are collectively more accurate and individually less expensive, cause no pain, present little radiation or other hazards, and have never been known to bring on death, as with angiography. It makes no sense to let anyone assault your body with catheters, dyes, and X-rays when innocuous vascular tests are available.

Noninvasive examinations may be highly sophisticated or exceedingly simple. Some come from space-age technology and others go back to Hippocrates (460–370 B.C.). There was an era when doctors, practicing the most elementary medicine and seeing black toes, would decide, "Yes, there is an impairment here!" Or they possibly looked at loss of hair growth on a leg and its generally poor color, whiteness, lack of blood supply, and came to similar conclusions.

In bygone years, the barbers acted as physicians. They used their five senses to check their patients and make diagnoses. As time passed, medicine progressed and trained physicians employed technical devices to tell them what was ailing their patients.

Over time, diagnostic instruments are adapted to new applications. It's not sufficient anymore, for instance, for a physician to take the blood pressure in only one of your arms. A proper noninvasive determination of the blood pressure throughout your body requires the doctor to check it in both arms and both legs. In fact, holistic physicians who practice true preventive medicine agree that if you don't have blood pressure readings recorded in both upper and lower limbs, you are not undergoing a complete physical examination. Why? Because there are medical parameters in which blood pressure of the legs should be equal to or slightly greater than that in the arms. If the leg pressures don't rise immediately following simple exercise, or if leg pressures drop after exercise, then there is an obstruction in your circulation. This occurs because when you exercise you increase demand and raise the pressure of the blood in the exercising limbs. Such measurements can reveal difficulties long before you get the usual leg cramps on walking fast, or the other typical signs and symptoms that usually come only after you are over 70 percent obstructed.

It used to be that the doctors felt coldness of the limb with their hands, until thermometers eventually were employed; except the thermometers didn't have sufficient skin contact and thus were not sensitive enough. Today, modern medical science has electronic thermisters for reading skin temperatures and infrared cameras that can measure and record temperature anywhere on the body without even coming near to the patient. Thermisters are currently used extensively for teaching biofeedback techniques. Ultratechnical devices such as those measuring body temperature accurately are now a normal part of an examination by many holistic and preventive medical practitioners.

THERMOGRAPHY FOR STROKE AND OTHER CARDIOVASCULAR DISEASES

When the United States came to exploring outer space, it was necessary to know the temperature of the moon before our astronauts could step onto its surface. In war, we invented the sniperscope for our servicemen to shoot accurately in total darkness; the device took advantage of the fact that the enemy solider's body hiding in the jungle was projecting heat and was warmer than the trees. The themographic infrared sensor in the sniperscope could easily detect the sniper, even at night. This sort of technology led to thermography, a new medical diagnostic technique using a camera to take appropriate infrared photographic views of the hands, face, feet, head, and other parts of your body.

The *thermograph* is a registering thermometer, one form of which records every variation of temperature by means of a stylus, moving with the mercury in the tube, and registering its rise and fall on a circular temperature chart turned by clockwork.

The *thermogram*, which the thermograph provides, is a picture that is a regional temperature map of the body or an organ. The picture is made without direct contact but is filmed by infrared sensing devices. Your radiant heat is measured, and thus the effective blood flow is recorded, when the environment is kept constant.

Thermography is useful as a screening device for stroke. It checks out potential stroke victims without subjecting them to any kind of trauma. You simply sit and have your picture taken; if it is not optimal, other appropriate tests can be done and, when warranted, medical prevention techniques such as chelation therapy may be administered long before the stroke occurs. In fact, you may have had no symptoms, or just loss of memory, or some dizziness—yet, significant obstruction of the blood vessels may be revealed.

Preventive measures are particularly important in the case of stroke. Even though the fatality rate when the cerebrovascular accident first hits is only about 15 percent, permanent disability is suffered by 50 percent of the survivors.[7] Of the estimated 2.5 million cases of post-stroke disability in the United States, one-third are wage earners, age thirty-five to sixty-five, who have become unemployable because of their stroke disability. The estimated annual cost of care for stroke disability in 1982 is $4.6 billion, aside from the costs incurred during the acute hospitalization. Thus the social, clinical, and economic rewards of detecting stroke before it strikes are tremendous.

Seventy-five percent of strokes in the forty-to-sixty age group are due to clogging or obstruction in the carotid arterial system. The carotid arteries are the two major blood vessels rising on each side of the neck to supply blood to the head.[8]

In the early narrowing stages of arteries about to bring on stroke before functional changes are noticed, the potential stroke victim experiences only an occasional and vague symptom known as transient ischemic attack (TIA). The TIA may consist of a momentary loss of vision in one or both eyes called a *fleeting blindness*, a speech loss that's temporary known as *expressive aphasia*, dizziness, memory loss, a sudden weakness of a limb that comes and goes quickly, or some other abnormality having to do with motor function or coordination in the brain. If you notice one of these symptoms from time to time, you may be receiving warnings that your arteries are undergoing hardening and stroke is liable to come on soon.

Thermography has been demonstrated repeatedly to be one of the several specific atraumatic diagnostic procedures that have proven to be useful in the detection and advance diagnosis of occlusive arterial disease such as the stroke I have been describing. Other noninvasive tests include *ophthalmodynamography*, the use of a device to measure the blood pressure in the ophthalmic artery. It is applied in conjunction with tonometric measurements of the eye and blood pressure measurements.[9,10]

Then there is *cervical auscultation*, a technique in which the blood flow sounds are picked up by a microphone placed over the common carotid artery. The sounds are correlated with the degree of arterial narrowing.[11,12] This phonoauscultation technique is useful for practically all parts of the arterial blood supply.

Funduscopy is an examination of the retina by an ophthalmologist or other skilled physician acquainted with the physiology of the eye. He or she is looking for evidence of hardening of the arteries, high blood pressure, and diabetes. Funduscopy may be permanently recorded with a retinal photograph.[13]

APPLICATIONS OF THE DOPPLER ULTRASOUND

About 15,000 American physicians are using ultrasound for the diagnosis of disorders in virtually every field of medicine, but especially in the cardiovascular system. The new technique is noninvasive and allows for two-dimensional motion studies. The use of ultrasound in diagnosis can be compared with doctors' viewing motion pictures of the heart or blood vessels rather than snapshots, which was the way ultrasound was adapted to medicine after World War II. Now the heart or blood vessels can be examined for an unlimited period of time; it's a "seeing" with sound.

The new technique adapts to the holistic concept, since it avoids the dangers of other specialized tests that involve insertion of needles and tubes and injection of chemicals that can cause potentially fatal allergic reactions. Ultrasound is free of radiation and has no known hazards.

Among its uses are the detection of brain tumors or head injuries, the spread of cancers through the body, gallstones, hidden birth defects, cirrhosis and other diseases of the liver, kidney cysts, and abscesses. In ophthalmology, ultrasound is used to help detect retinal detachments, tumors, metal fragments, and other foreign bodies that threaten to blind when they lodge in an eye. It has been credited in other medical problems with decreasing the length of hospitalization and making the diagnosis earlier than would have been the case with other tests. The ultrasound equipment is simple and portable and is able to be brought to the patient's bedside.

In general, the *Doppler ultrasound device* relies on an echoing principle whereby sonar signals that are too high to be heard by the human ear are transmitted, bounced off a target, and then received by the source, a transducer. This small instrument, made as small and handy as a pencil, contains a piezoelectric crystal that vibrates when charged with electricity and changes voltage into sound. The crystal can also pick up the reflected sound that it has generated and change it back to electricity.

In the two-dimensional motion studies supplied by larger Doppler ultrasound devices, doctors don't rely on just one beam. They use a combination of up to thirty beams or a single beam in sweeping fashion to produce images at a rate of thirty each second, for viewing as a motion picture.[14]

Echoarteriography of the carotid blood vessels is an ultrasound reflection technique to note the presence or absence of disorders. There is *echoencephalography*, the investigation of structures within the skull by detecting the echoes of ultrasonic pulses. And in *echocardiography*, ultrasound waves investigate and display the action of the heart as it beats.

THE WORKINGS OF PLETHYSMOGRAPHY

Another highly sensitive diagnostic instrument is the *plethysmograph*, which is a device for measuring and recording changes in blood volume of a body part, organ, or whole body. The plethysmograph is a blood-pressure-like apparatus applied to the finger, toe, ankle, wrist, or other body part. The body part simply lies at rest on a table. The plethysmograph cup is applied to the part and measurement is made of the amplitude of pulse volume. The measurement tells the status of your circulatory system and whether it is in spasm or blocked or healthy.

You may be subjected to a brief exercise period on a *treadmill*, because any muscle that is exercised should have increased blood pressure and blood flow, as I mentioned earlier. If no increased blood flow is demonstrated on the plethysmograph, this indicates you may have advanced narrowing of one of your arteries.

Plethysmography measures the blood volume and muscle tone of an artery. Characteristics of a healthy artery are similar to those of a new automobile radiator hose. The hose shows softness, pliability, springiness, strength, stretchability, smoothness, and durability. An artery with these same characteristics has a great influence on the speed and volume of blood flow. When the heart pumps, its blood is delivered throughout the body virtually in an instant. The artery stretches to let a bolus of blood go forward and immediately closes behind the bolus and thus helps push it on to the next point.

A soft set of blood vessels takes half the work load off the heart. Your healthy stretchable artery acts as an auxiliary pump by opening easily to receive the blood bolus without having to be forced. Then its pliant, muscular wall clamps down behind the bolus of blood to pump it along.

The plethysmography technique tells your doctor the amplitude of blood volume in an artery and how fast the blood bolus moves past the particular anatomical part being measured. Plethysmography is a totally nonhazardous, noninvasive examination. It is exceedingly accurate and useful to determine the circulatory status in your arms and legs, and even the brain and eyes can be tested.

THE EXERCISE TOLERANCE ELECTROCARDIOGRAM

A record is made by the electrocardiogram (EKG). The EKG is an apparatus universally available to physicians and has been part of routine physical examinations for years. As well as showing other cardiovascular abnormalities, the EKG is supposed to tell of injury to the heart muscle by some specific indication of a "leak" in the EKG battery whereby electrical potential escapes even during the period of muscular rest. In such circumstances electrical potential is never zero, and the baseline (starting point) shifts away from its usual zero point.

In normal individuals the electrocardiogram shows no change except for a faster heart rate after exercise. In those with an inadequate blood supply to the heart, specific changes may sometimes be observed. But in truth, the resting EKG is of little value in revealing a lack of blood supply to the heart. An EKG should be taken only when the heart is stressed.

Provocative or stress techniques have been used effectively in many areas of medicine. The classical example of a worthwhile stress technique is the exercise tolerance electrocardiogram. Such a noninvasive stress test measures the maximum effort you can expend without overstraining the heart. This is done with tests on an exercise bicycle or a treadmill under the supervision of a physician.

Sensors are attached to your body to monitor heart rate. A rubber cuff for blood pressure readings get attached to your arm, as well. You pedal the bicycle at gradually increasing speed for about twenty minutes. The physician observes the EKG for readings of the heart's electrical action in response to the growing stress. Periodically he takes blood pressure readings to monitor the intensified force of your heartbeat and how efficiently it continues to pump the blood through the arteries.

Your rate of blood flow, the elasticity of the main arteries, which controls steadiness of blood flow, resistance to flow within the arteries, and the thickness and quantity of your blood are all factors influencing blood pressure. In turn, these affect the heart's stability as the pump of the body's circulatory system. The same procedure may be followed using a treadmill instead of an exercise bicycle.

Graduated stress tests are also being used in a series to learn the factors that may trigger attacks of angina pectoris in individuals. As stress increases, the heart requires more of its own blood supply from the coronary arteries to provide extra oxygen. When these arteries nourishing the heart are obstructed or narrowed from plaque buildup, they can't provide enough blood flow to answer the heart's demand and the pain of angina occurs. For an individual with this coronary artery limitation, the degree of stress at which pain occurs during the testing is the level of activity to which the person should restrict himself.

Unfortunately, there are significant limitations to the value of exercise EKG for diagnosing coronary disease. Women have been found to have a high proportion of false positives (54 percent) while men only had a 12 percent false-positive rate. Interestingly, men had the disease much more commonly than women—70 percent versus 29 percent; however, after a documented heart attack, the presence of S-T segment depression is highly indicative of multivessel disease. The stress EKG is an important and highly useful test and must be done, but it provides probabilities more than absolute diagnostic certainty.[15] Improvement in exercise EKG following chelation or other therapy is important, highly desirable, and, fortunately, quite common.

THE FIRST-PASS RADIONUCLIDE ANGIOCARDIOGRAM

The transit of radiotracer coronary blood pool during its first entrance through the right heart, out through the lungs, and back through the left ventricle is measured by the *radionuclide angiocardiogram*. To establish the resting condition of the heart, the tracer is first injected at rest. The patient is then exercised on a bicycle ergometer to achieve a high heart rate of 80 or 90 percent of his predicted maximum ability. A second dose of the tracer is injected at this elevated heart rate to measure if the patient has any heart or artery insufficiency due to the great oxygen consumption of the heart muscle.

This exercise technique increases the work load of the heart from five to fifteen liters of blood per minute. If any underlying narrowing of the coronary arteries from atherosclerosis exists, the investigating nuclear cardiologist will see it right away. The radionuclide angiocardiogram is far superior to any EKG, either resting or of the exercise tolerance type. Of course, the radionuclide test also has the potential of eliminating angiography from ever being considered again.

The EKG may show the absence of pathology, but the radionuclide angiocardiogram will still pick up heart problems far in advance of their showing on the current standard tests. Even in the face of a normal stress electrocardiogram, the newer technique can demonstrate if you have underlying heart or coronary artery disease.

The tracer is technetium protecnitate, with a half-life of six hours, the same tracer that has been used safely for years for brain scans. It's the principal isotope employed in nuclear medicine, injected intravenously and excreted through the intestines. The actual measurement of the tracer bolus takes only about seven seconds, which is the usual transit time of a person with a pulse of 72 beats per minute. A special camera, the Cordis Baird System 77, or the scintillation camera manufactured by Searle Radiographics, or some other similar system, takes high-speed pictures at exceedingly short intervals of time of the blood coursing through the cardiovascular system.

There is no X-radiation involved. The gamma radiation present is 140 KEV photon, equivalent to one-tenth of what you would receive from one chest X-ray exposure. It's easily accomplished safely for children (even every day for a year). A study of the heart using this system costs from $300 to $800, and a study of the head for brain circulation runs $200 to $400.

HAIR MINERAL ANALYSIS

Base-line medical testing of the general arterial circulation will usually be the way a person with poor arterial function finds out he or she has problems. Physicians who practice pure preventive medicine (the holistic concept of health) will test for *blood lipids, glucose tolerance,* and *mineral status,* and perform *diet analysis* and *hair mineral analysis*. Unfortunately, most physicians do not include these particular tests in their laboratory evaluations. That being the case, as an informed patient you must ask for them yourself.

Hair analysis, for example, has been known for years to be valuable in identifying subclinical signs of chronic mineral deficiency or mineral excess even before actual overt disease has developed. It's a useful way of screening patients and helps to detect subclinical illness before any serious signs or symptoms make their appearance. When used with some of the other tests described here, hair mineral analysis provides accurate data that is not readily available by any alternative method.

Unlike blood or urine tests, hair analysis reflects the body's levels of mineral content inside cells and reflects long-term excessive exposure and/or potential retention of heavy metals.

As a tissue specimen of your body, the hair has several advantages. It is biologically stable, stores easily, ships well to a distant laboratory for the most economical and efficient analysis, and will not deteriorate. Hair is easy to obtain. Clipping it is nontraumatic and noninvasive. Unlike blood and urine, hair doesn't get altered in the analytical procedure by chemical changes.[16] It reflects your actual systemic intracellular mineral levels.[17,18] Intracellular minerals are usually quite different from levels seen in blood or urine, because intracellular levels are very different from extracellular mineral levels. Hair reflects the average mineral intake or exposure over the past one to two months, whereas blood and urine reflect the values only at the moment when they are collected. The hair is more easily and accurately measured than other potential body specimens reflecting intracellular levels, such as bone or liver tissue. Hair is in equilibrium with systemic activities and indicates your metabolic balance over the period of one to two months or longer, depending on the length of the speciman of hair being analyzed. In brief, hair analysis is a unique capsule indicator of your body's general condition and not just a test of one tissue or organ.

The medical profession knows that the majority of diseases begin with biochemical imbalances occurring at the cellular level. As you learned in Chapter 3, hardening of the arteries starts with the spontaneous subdivision of just one smooth-muscle cell in the media of an artery. "Deficiencies or excesses of metal ions cause most adverse biological effects," says an important textbook describing the newest physical science, known as bioinorganic chemistry.[19] In the case of atherosclerosis, the excess of metal ions comes from an aberration of calcium metabolism.

In an editorial of the *Journal of the AMA* the statement "We are becoming, however, increasingly appreciative of our dependence on trace elements. We are truly at their mercy" demonstrates the deepening recognition being given by scientists to the importance of trace elements in health maintenance. Research has been pointing out that modern health care has the potential to correct metallic ion imbalances before disease reaches the stage of physical symptoms.

Trace minerals are building blocks for the biochemical processes of life. As vital components of all essential enzymes, they are necessary for overall good health. Essential element imbalance or interference due to heavy metal accumulation will lead to your body's chemical imbalance and leave the door open to potential disease, especially cardiovascular disease. I will discuss heavy metal toxicity at length in Chapter 8.

Hair mineral analysis is a way to monitor your body for its trace mineral content. Orthodox physicians have utilized blood, urine, spinal fluid, tissue biopsy, even feces and sputum to evaluate the body's health status. It's time they routinely added hair to specimens being analyzed for the useful data they provide. Each method has strengths and limitations, and none should be relied upon as a sole trace mineral diagnostic tool. But, as an overall health indicator and as a measuring tool for biochemical mineral balance, hair analysis is an "ideal method" to monitor your exposure to heavy metals that cause disease.[20]

All it takes to get your hair analyzed for trace elements is a couple of tablespoonfuls of hair from the nape of the neck, which would be sent by your doctor to a licensed hair analysis laboratory, specializing in this work, such as Mineralab, Inc., Garry F. Gordon, M.D., Medical Director, 3501 Breakwater Avenue, Hayward, CA 94545; telephone (415) 783-5622. If your physician is narrow in his outlook or resistant to adopting a medical method with which he is unfamiliar, do it yourself. The International Foundation for Health Research will assist you. Write P. O. Box 5026, Hayward, CA 94540. You can also get hair mineral analysis from the International Foundation for Preventive Medicine, 7430 Mason Lane, Falls Church, VA 22042; telephone (703) 573-2500.

Noninvasive diagnostic testing is a lot more valuable for the patient and the examining physician than any invasive test such as angiography. Noninvasive tests tell you what is really happening physiologically. They are well accepted by even the most conservatively minded physician. In comparison, the invasive test (arteriography) I've discussed is strictly anatomic—providing static information—and may indicate that a blood vessel is pumping adequately when all the time it is a rigid pipe. Or the artery may appear to be 100 percent closed on the angiogram when at the moment the X-ray was taken it actually was only in temporary spasm. The angiogram gives a misleading conclusion, while the noninvasive tests can tell how the circulation is functioning under action conditions.

CHAPTER FIVE

The History of Chelation Therapy and Why It Is Not Commonly Available

EDTA chelation therapy has been used around the world for almost half a century as the definitive treatment for cardiovascular disease; just the dosage of EDTA has changed over the years. Nevertheless, only recently has the orthodox medical community in the United States begun to recognize the treatment as a preventive or reversal agent for hardening of the arteries. Don't be surprised, therefore, if your family doctor isn't enthusiastic when you mention this procedure. The clinical articles in the medical literature may have escaped the physician's notice. After all, there is an awesome amount to know in the practice of medicine. Some physicians tend to be down on even established medical techniques that they aren't familiar with. Perhaps you'll bring this book's bibliography to your doctor's attention so that its medical references can be checked. Why not give the physician a chance to go to the medical library and read what his or her colleagues have written about chelation therapy?

As a potential user of this life-extending treatment, you should be aware that chelation is as old as life on this planet. Many biological processes—in nature and in mankind—are involved in the phenomenon of chelation. In Chapter 7 I shall explain how and why this is true. It is only in recent years, however, that EDTA chelation has emerged as a very valuable therapeutic technique for purposes of predictable health improvement.

THE HISTORY OF EDTA CHELATION

EDTA was first synthesized by Franz Munz, a German chemist, at Hochst, Farbwerke, Frankfurt, for nonmedical purposes, and developed by the I.G. Farben Industries as a result of the economic problems created by Adolf Hitler. Germany required the product in its fabrics and textiles industries. EDTA was patented in 1930, as reported in the *American Journal of Laboratory and Clinical Medicine* ("Regarding EDTA," 1930). Working with some older components, I.G. Farben finally brought out the agent in 1931 to substitute for citric acid as a coating or preservative for cloth fibers.

Chelation therapy was first employed in medicine in 1941 with the use of sodium citrate in lead poisoning. This use was reported in an article, "Treatment of lead poisoning with sodium citrate," by S. S. Kety and T. V. Letonoff (*Proceedings of the Society of Experimental Biological Medicine,* 46 [1941], 476–77).

EDTA was patented in the United States in 1941 by Frederick Bersworth, a biochemist at Georgetown University, for the Martin-Dennis Company. Several papers on EDTA were published in the early 1950s.

EDTA is known to most chemists. There are entire books devoted to it. Published articles dealing with EDTA in the field of analytical chemistry number more than 5000. One of the values of this agent is its precise nature. In the test tube, a scientist can mathematically predict how EDTA is going to act and approximately to what extent it is going to chelate. It is used extensively by the food industry as a stabilizing additive for canned goods, salad dressings, and other products. Read the label on your jar of Hellmann's "Real" Mayonnaise, for example, and see that you've been eating EDTA for years.

In a March 1976 appearance before the Advisory Panel on Internal Medicine of the Scientific Board of the California Medical Association (CMA), Norman E. Clarke, Sr., M.D., a cardiologist from Birmingham, Michigan, considered the "grandfather" of this treatment in the United States, told the Scientific Board of his pioneering activities with chelation therapy.

"I learned about EDTA in 1953 from Dr. Albert J. Boyle, Professor of Chemistry at Wayne State University, Detroit, and from Dr. Gordon B. Myers, who then was Professor of Medicine at the same university and a well-known cardiologist. Drs. Boyle and Myers had preliminary experience in treating two patients at University Hospital, Detroit, who had calcified mitral [heart] valves. The patients were almost completely incapacitated," Dr. Clarke told the CMA. "The doctors were pleased with the results [of chelation treatment] because they obtained very satisfactory return of cardiac function. But they did not have the opportunity or time to go on with it [their research]. Dr. Boyle asked if I would—because then I was chairman of the research department of Providence Hospital, Detroit—undertake a study of EDTA in cardiovascular diseases, which I did.

"I knew, having been in cardiology quite a number of years [since 1921], that arteriosclerotic cardiovascular disease was a helpless, hopeless situation for the cardiologist. I had to start by trying to find out whether it [EDTA] was safe; what was the best dosage; were there any side effects; and how would we standardize its use as a treatment?

"Well, of course," Dr. Clarke continued, "the first couple of years we treated only hospital patients where we had them under excellent control. We started with rather large dosage [10 gm] and had some side reactions consisting primarily of signs of a [vitamin] B_6 deficiency. Some male scrotums lost a complete cast of skin, but it was absolutely painless and with no sensation of discomfort. After finding the proper dosage, of course, all those things have never happened again. And even with those large doses we had no unusual or serious side effects. Ultimately we determined, and for a long time gave, 5 grams, and later determined that 3 grams was the proper dose.

"In the last twenty-eight years of my experience with EDTA chelation I would say conservatively, because after all those years you don't keep accurate records with all you do, but conservatively I have given at least 100,000 to 120,000 infusions of EDTA and seen nobody harmed," the chelation specialist said. "I've never seen any serious toxicity whatsoever. I've seen only benefits. . . .

"I early observed its superior beneficial effects in vascular diseases of diabetics," Dr. Clarke added. "In one of my early satisfying results a thirty-eight-year-old man was brought into the hospital suffering from extremely severe leg rest pain [pain with no movement]. It was very severe. He had gangrene of his toes. His was an emergency. We decided on a special method [of treatment]. We put a catheter by needle into the femoral artery of the leg where the trouble was, and devised a way of keeping a twenty-four-hour [EDTA] drip. Within twenty-four hours his pain was relieved with very satisfactory results. And later on, within forty-eight hours, his gangrene was improved. That was an immediate result of Endrate [EDTA chelation] therapy, delivered directly into the artery of the involved leg."

The veteran cardiologist went on to describe more success with diabetes and other cases of rest pain. Then Dr. Clarke said, "Another field in which I have found [EDTA] most effective is cerebrovascular senility. I am extremely impressed with that, not only by the improvement in the patients but from the economic factor today. All you have to do is go into one of those old folks' homes and see the senile people sitting around and you'll recognize immediately not only the economic problems but the physical and mental problems that they and their families go through."

Dr. Clarke, who is over ninety years old himself and stays alert and youthful by giving himself EDTA intravenous injections, concluded his remarks to the CMA advisory panel assembled to hear about chelation therapy, saying: "After all these years and with all that experience, I am just as certain as can be that EDTA chelation therapy is the best treatment that has ever been brought out for occlusive vascular disease."

Despite Dr. Clarke's long successful experience using EDTA, the organized medical profession in the U.S. has continued to reject it until this time.

ORIGINAL USE OF EDTA FOR HEAVY METAL TOXICITY

The original application of disodium ethylene diamine tetraacetic acid in the United States was for the treatment of heavy metal toxicity. In this country EDTA was first used against lead poisoning in workers employed by a battery factory. Lead, a toxic heavy metal, was removed from the bloodstream and other body storage areas of workers by means of intravenous infusion.

EDTA chelation was also of interest to the U.S. Navy for removing the lead from sailors who absorbed it while painting ships and dock facilities with lead-based paint. The therapy is still used for that purpose.

The medical potential of the chelating agent was demonstrated dramatically in 1951, when EDTA saved the life of a child suffering from lead poisoning (see "Chelation" by Harold F. Walton, *Scientific American*, June 1953).

Physicians observed that patients who had both lead poisoning and hardening of the arteries began to improve greatly following chelation treatments with EDTA. Seeing their vascular conditions change for the better, the doctors started to treat others with artery-hardening conditions who did not have lead poisoning. It was no surprise that patients with just hardening of the arteries improved, too. In fact, the investigation of the medical uses of chelating agents has produced a voluminous literature, and chelating drugs have been developed for the treatment of a wide range of diseases from metal poisoning to cancer. The latest use of chelators is for treatment of radioactive isotope poisoning.

Besides citric acid and aspirin, other common agents that work, at least partially, through the chelation mechanism are cortisone, Terramycin, and Adrenalin. A host of additional chelating compounds, natural and synthetic, exist—probably numbering in the tens of thousands. In Chapter 7 I describe some you can employ yourself at home.

The Food and Drug Administration (FDA) at one time allowed the pharmaceutical companies to state that EDTA was "possibly effective" in occlusive vascular disease. This ligand (a molecule bound to a central atom) never achieved "completely effective" status for hardening of the arteries, because the pharmaceutical companies decided against spending the necessary funds to get this status. Today, the FDA has removed "possibly effective" from the EDTA package insert.

Intravenous infusion is the only administration route used, for the chelating substance is poorly absorbed and not well tolerated when taken by mouth. It does not cross the intestinal membrane well, with only approximately 5 percent absorbed on the average. Your stomach gets upset when you swallow EDTA in the required higher dose for any potential beneficial therapeutic effects. Even so, the ligand is a common additive to many foods, as I mentioned. In this form, as a food additive, it cannot help your circulation.

Also, orally ingested EDTA doesn't cross cell membranes well. That's one reason why EDTA tablets are no longer prescribed for a detoxifying therapeutic effect. In fact, oral administration of some forms of EDTA in therapeutic dosages may simply enhance heavy metal toxicity. The orally administered EDTA tablet may cause the lead, mercury, cadmium, aluminum, or other divalent and trivalent toxic heavy metals (which we regularly consume without realizing it from our air, water, and food) to cross the lining of the digestive tract and enter into your body; otherwise, they might simply have passed through the bowel and out. Does EDTA do this when you eat it in salad dressing? Possibly, if you saturate your salad with large quantities of dressing, you'll get less nutritional value from the salad. Generally, however, in food the EDTA is already bound to the useful trace minerals such as zinc and iron. You simply will get less minerals from these EDTA-coated foods. The American Academy of Medical Preventics considers it unethical for one of its physician members to give patients tablets containing EDTA for the treatment of arteriosclerosis.

Chelation with EDTA administered intravenously remains an FDA-approved method for the emergency treatment of hypercalcemia and also for routine lead detoxification from the human body. It is excellent for these purposes. This was confirmed as far back as 1961 during the Federation Proceedings on the Biological Aspects of Metal Binding.

The State University of New York, Downstate Medical Center College of Medicine in Brooklyn, recently used calcium EDTA to treat hyperactivity of children, with superb results. The medical authorities there believe that these children may have benefited because their hyperactivity probably represented an increased susceptibility to lead toxicity. It's likely they did not have true "lead poisoning" by the older accepted definition. Their bodies were rendered lead-toxic possibly due to a diet low in calcium, zinc, vitamin C, protein, or fiber. These nutrients give vital protection against lead.

Hyperactivity in children suffering from minimal brain dysfunction is a sign of possible lead toxicity.[1,2] Additionally, myocarditis,[3] neuropathy due to lead,[4] and hypertension due to cadmium are among those heavy metal toxicity diseases now recognized. Heavy metals also cause suppression of our immune system, so we are more vulnerable to everything from "colds" to cancer to neurotoxicity, causing us to forget things too easily. Chapter 8 contains information on these subjects.

After chelation, follow-up hair analysis generally confirms the decrease in heavy metals as well as the improved trace mineral levels (when adequate nutritional supplements have been taken). Failure of the patient to show health improvements may suggest the need for further in-depth evaluation by the chelating physician. He or she will probably look for some continued but still unrecognized, source of heavy metal exposure, such as the glazing on your dinner plates or on drinking glasses or in drinking water. The doctor checks for potential contributing absorption or excretion defects in trace mineral metabolism. Hair analysis, which I described in Chapter 4, is a very valuable test to enhance the effectiveness and safety of chelation therapy.

Zinc and chromium deficiencies are readily diagnosed with hair testing in order to help eliminate all potential aggravating causes of arteriosclerosis,[5,6] as well as to help maintain chelation benefits, and to achieve even greater improvement with the maximum elimination of symptoms. Hair analysis also contributes to greater chelation safety by avoiding potential aggravation of mineral deficiencies (such as zinc and chromium) through early recognition of these increased needs in some patients. Today's chelating physicians minimize the potential for EDTA toxicity by providing appropriate mineral and vitamin therapy along with the intravenous injections.[7-9]

VITAMIN AND MINERAL THERAPY IS NEEDED

Chelation therapy is now known to bind and thus remove certain of the B-complex vitamins,[8] particularly pyridoxine, as well as many essential trace metals,[9-11] notably zinc and chromium. It is therefore logical that clinicians using chelation therapy had to become keenly aware of the recent developments regarding the benefits of these vitamins and minerals in preventing hardening of the arteries.[12,13] Mineral therapy and vitamin therapy are integral additions to the EDTA chelation injection program. Hair analysis for evaluation of the patient's trace mineral status is performed routinely.[14-17] Obviously, then, the program of treatment against hardening of the arteries is not confined to EDTA chelation therapy alone. The anti-atherosclerotic program includes suggestions for dietary supplementation, vitamins, chelated minerals, special foods, exercise, and other items.

Soviet physicians employed a chelator called Unithiol with a multivitamin administration in coronary arteriosclerosis. The Russians encourage the use of vitamin and mineral supplements for their people. They cited several sources in their 1973 and 1974 clinical journal articles regarding the use of vitamins and Unithiol in arteriosclerosis.[18-20] The Soviet medical scientists concluded that such early treatment with dietary supplements and chelation is the most successful anticoronary approach. They recommend this combination treatment for the prevention of hardening of the arteries in the aging as well.

This comprehensive approach of chelation and mineral and vitamin therapies may explain the great success some of today's clinicians are seeing in treating vascular disease. It is truly a "holistic" approach (whole body healing). Other informed and forward-looking doctors are beginning to use these holistic concepts as a more total treatment program for their patients' health problems.

The Soviet chelating agent that I mentioned, Unithiol, employs a sulfhydryl group as the chelator.[19,20] When combined with orthomolecular nutrition (the use of megadose nutritional supplements) and regular exercise, this approach was so successful that some knowledgeable Russian physicians now use it as an antiaging procedure. It turns back the clock on the wearing out of tissues and reduces the incidence of hardening of the arteries for many older people.[20] A similar program is employed in Czechoslovakia. In these two countries, EDTA chelation is a commonly used treatment against all kinds of diseases derived from hardening of the arteries.

In 1972, the Czechoslovakian article "Chelates in the Treatment of Occlusive Atherosclerosis"[20] concluded that EDTA was the treatment of choice for vascular disease producing intermittent claudication of the legs.

It should thus be clear that all of us in the Western world have at least four choices by which to govern how and when we might live and die. Until now, only two of these choices have been offered. The choices are: (1) the extreme of simply watching and waiting for death caused by hardening of the arteries; (2) the other extreme of traditional medicine's totally inadequate "medical" *palliation* for the relief of hardened artery symptoms in whatever organ the disease happens to strike;[21] (3) the intermediate between these of vascular surgery as a life-preserving technique, which consists of rather radical surgical procedures; or (4) what this book shows you as a fourth choice, the preventive and reversal treatment with EDTA chelation and nutritional therapies.[22] Which do you choose?

HOW TO INCREASE AMERICAN LIFE EXPECTANCY

Chelation therapy allows a person with occlusive vascular disease from hardening of the arteries and arterial spasm to optimize his response to all the various forms of therapy he takes. As a single-modality approach, EDTA chelation is unsurpassed.It must be combined, however, with a comprehensive, multitherapeutic approach that includes reestablishment of optimum metabolic equilibrium through diet, exercise, reduced stress, and nutritional supplementation. Then the patient experiences an unexcelled effect. This is the orthomolecular-holistic approach, by which occlusive vascular disease processes will be generally controlled and possibly reversed.[22-34]

Too bad that even today this treatment, with its minimum of imperfections and maximum of benefits, has been held back from Americans by practitioners of mainstream medicine. Still, we could change the findings of the World Health Organization, which declared the United States to be the only nation in the world that had shown virtually no increase in life expectancy in recent years. The United States now ranks seventeenth among nations in life expectancy.

All known risk factors for hardening of the arteries must be identified, treated, and eliminated, if possible. For example, defective lipid metabolism must be improved with diet, nutritional supplements, and hormones, instead of the current inadequate practice of surgical expediency or the other extreme of watching and waiting. Look at the prevalent practice of American medicine as currently carried out—we wait for a crisis after watching a disease pattern take hold. Then we attack the problem—not infrequently using expensive and dangerous surgical procedures.

Present forms of orthodox American crisis-oriented medicine for hardening of the arteries invariably arrives with too little, too late. And chelation or supplemental nutrition is not even included as one of these orthodox therapies. I believe hardening of the arteries and other forms of chronic, degenerative diseases that confront Americans can be reduced and even eliminated through application of these orthomolecular-holistic medical concepts, which is the practice of preventive medicine in its most ideal form.

The intravenous approach for the administration of EDTA has been made a part of the holistic approach. However, this technique to increase every American's life expectancy no doubt could be improved upon. Someday, perhaps, one infusion might do the work of several. This improvement would lower the cost associated with current administration techniques and thus increase the acceptance of the concept, not only for the restoration of blood flow through diseased vessels, but also for the amelioration and prevention of several other disease processes. Until that general acceptance by the people and their physicians occurs, the current availability, components, administration technique, and costs of intravenous chelation are not likely to change much. My fear is that acceptance may come too late to have this important therapy included for reimbursement in any forthcoming national health insurance program. If that were the case, its availability to the American people could be seriously threatened by the ever-present red tape of bureaucratic regulations usually associated with government-controlled medicine. As we shall see in Chapter 10, Medicare is presently denying coverage to that part of the U.S. population desperately needing chelation treatment.

TWENTY REASONS AMERICAN ORTHODOX MEDICINE HASN'T PRESCRIBED CHELATION THERAPY

With nearly thirty years elapsing since chelation therapy was reported beneficial in reversing hardening of the arteries, it still is not generally accepted by orthodox medicine. The use of this treatment remains relatively unknown and controversial in the United States, although a number of countries such as Russia, West Germany, Czechoslovakia, Japan, Norway, Sweden, Yugoslavia, and others have begun to make it their first choice for rejuvenating the cardiovascular system. It's logical for you to ask why such a circumstance exists in our own country. I've wondered as well. In studying the history of chelation therapy, the participating physicians of the American Academy of Medical Preventics acknowledge twenty reasons that could be responsible.

The average local physician who practices in the usual orthodox style should not be blamed for taking a dubious attitude toward the treatment. For over three decades, doctors have been exposed to the twenty relevant factors in the shaping of their medical opinion about vascular disease and its therapies. Keep in mind that these factors focus on chelation therapy as the fall guy in the present circumstance, but many other excellent therapeutic agents fit into similar molds. Such agents are martyrs to the American medical system. And who are the final victims? They are those Americans suffering with serious cardiovascular illness who need these various therapeutic agents so desperately.

The average local orthodox physician is influenced away from choosing the chelation answer for his or her patients who have hardening of the arteries by the following twenty factors:

1) During the early times when EDTA was first injected for reversing heart and blood vessel diseases, the medical researchers were still looking for a one-shot "magic bullet" to do the job. They failed to take into account that specific lifestyle improvements were required by the patient whose arteries had deteriorated over many years. The doctors permitted patients to continue smoking, get no exercise, eat a high-fat and high-sugar diet, take no nutritional supplements, and do other damaging things to their bodies. No comprehensive therapeutic approach was employed, as it is today, to minimize the potential for unsuccessful treatment of vascular disease. Consequently, there were failures, some of which were at least partially due to continued heavy smoking and other defective lifestyle problems such as improper diets, mineral and vitamin deficiencies, or other nutritional imbalances. All of these failures were entirely blamed on EDTA chelation therapy.

In spite of the early researchers witnessing initial apparent excellent responses to the intravenous EDTA infusions, when no concurrent supportive dietary and exercise changes were made in the patients' lives, many of these patients eventually succumbed some years later. Autopsies revealed that hardening of the arteries was still present. The chelation treatment had not removed it all—forever—as the early researchers had hoped it would. These failures were talked about in the medical community and cast a still lingering cloud of suspicion on the treatment's efficacy.

2) Only recently has hardening of the arteries been understood to come from a combination of complex causes, including the repeated abnormal subdivision of a smooth muscle cell, aberrant calcium metabolism, binding of this calcium with lipids and polysaccharides and other substances into a cementlike material, cross-linking of collagen and elastin, platelet aggregations, and vasospasm. Medicine's new comprehension of the complexity of arterial disease has finally caused doctors to abandon their search for a single therapeutic "magic bullet." They finally are also beginning to understand that the complex causes take years to develop and are reversible processes that, therefore, can also be largely prevented. Chelation therapy is

one of the major tools in bringing about prevention or reversal, but the news is just beginning to spread throughout the medical profession.

3) Medicine and the media have grossly overemphasized the simplistic blood fat theory of hardening of the arteries. Orthodox physicians tend to approach the problem of narrowed blood vessels as if it was an obstruction of inert tubes. Arteries and veins, of course, are comprised of living, pumping, flexible cells. Chelation therapy does not fit well into the fat theory concept. It's an erroneous and misleading informational context. Although intravenous EDTA does reduce serum cholesterol levels and improve liver function and copper-to-zinc ratios, these mechanisms are not sufficient to explain the remarkable clinical results seen in chelated patients with impared circulation. The treatment simply doesn't fit the old atherosclerosis model that doctors were taught. Medicine's current interest in aberrations of calcium metabolism as it relates to vasospasm and the widespread interest in calcium antagonists has profoundly altered doctors' understanding of occlusive vascular disease. Eventually intravenous EDTA injections may finally receive the attention this treatment deserves when calcium channel blockers become more popularly employed, and the limitations and side effects of the newly approved calcium blockers begin to be appreciated.

4) Current advances in bioengineering have produced significant developments in noninvasive blood vessels and circulation testing. Use of some of this new technology is easily accomplished in any physician's office if he or she makes a financial investment to acquire the equipment and takes time to learn about it. Painless, noninvasive tests will predictably eventually replace the expensive and dangerous angiograms performed today in hospitals, which all too often lead to coronary artery bypass operations. The primary reason angiograms are performed are as diagnostic precursors to open heart surgery. Instead of vascular surgery, physicians performing in-office noninvasive tests find themselves more inclined to look for other nonsurgical techniques to improve circulation for their patients. But noninvasive testing equipment is just becoming widely accepted in the medical community. Most doctors have not yet had the opportunity to explore in depth the new medical and holistic, orthomolecular alternatives to heart and blood vessel surgery.

5) Angiography has unfortunately been relied on for over thirty years to provide a supposedly accurate diagnosis and then to determine the success or failure of any treatment. As I reported in the last chapter, now we have proof of just how erroneous these readings of angiograms have been, even when done by the most renowned angiologists in the field. They simply have not been giving correct, reproducible information. Yet, surgeons are still operating based on these imprecise studies—and patients are dying from operations they did not need!

We finally know why before-and-after angiographic studies following an apparently successful therapeutic regimen with excellent clinical results using chelation therapy may fail to be reported by angiologists as significant visible improvements. Their angiograms may even allow the patients' arteries to look worse when dramatic clinical improvement may readily be documented by noninvasive vascular tests. Clinical improvement is explained by one or a combination of several physiological mechanisms such as increased collateral blood flow, decreased blood viscosity, improved cell membrane functioning, better organelle (a specialized cell structure) functioning, decreased vasospasm, and others. An angiogram cannot measure such functional or physiologic improvements, since it merely shows the anatomy of arteries and not their physiology. We may finally be getting rid of our excessive reliance on angiography, which will allow medicine to concentrate on more useful and accurate tests that lead to better treatment.

6) By training almost 1000 catheter-oriented specialists each year, American medicine is forcing itself to favor invasive techniques of diagnosis and surgery. Cardiologists have a vested interest in continuing the use of invasive coronary angiographic procedures far in excess of our needs. Cardiovascular surgeons have proliferated in number, as well. Neither group of specialists has any interest in promoting the use of noninvasive therapies that could be performed in-office by the average family physician. Unfortunately for the patient, these two types of specialists have a great deal of control over the direction that cardiovascular treatment must follow—to the extent that illegal-restraint-of-trade litigation may have to be undertaken against them (see Chapter 11).

7) Ignorance, hostility, and misinformation from cardiovascular surgeons, peripheral vascular surgeons, and other specialists whose income is gained from the employment of invasive studies and traditional therapies, have kept patients from knowing enough to bother about investigating chelation therapy as a viable alternative. Laypersons are frequently given completely false information regarding chelation therapy. The best estimate is that heart surgeons and other vascular specialists will lose 75 to 80 percent of their referrals, since most chelation patients are restored to a pain-free and active lifestyle. Although not all of these patients achieve significant reversal of their arteriosclerosis, three-quarters of them will have improved circulation sufficient to go back to an active symptom-free life.

8) It's only within the past couple of years that chelation clinicians have been able to document their cases, using the new, sophisticated, and accurate radioisotope studies. Until now, the refusal of these holistic doctors to subject their patients to dangerous before-and-after angiography studies has contributed to the lack of recently published information in the scientific literature regarding the dramatic clinical improvements seen in

arteriosclerotic heart and blood vessel diseases in patients receiving chelation therapy. But with the recent development and broad acceptance of noninvasive vascular studies, we now find at least three reported recent investigations documenting excellent improvement in blood flow among chelated patients.

The Baird System 77 Multicrystal Computerized Radioisotope Camera was first employed by chelating psychiatrist Lloyd Grumbles, M.D., of Philadelphia on chelation patients. These studies, presented as a lecture to the AAMP in 1979, revealed amazing before-and-after improvement in cerebral blood flow in over 80 percent of Dr. Grumbles's chelated patients.

Also, using ultrasensitive radioisotope studies performed in the Neurosurgical Department at Loma Linda University, Bruce Halstead, M.D., of Loma Linda, California, consultant to the World Health Organization and present chairman of the AAMP board of directors, documented in lectures at two AAMP science conferences in 1980 and 1981 significant improvement in circulation to both the brain and the heart in a dozen of his patients taking a course of chelation therapy.

And cardiologist H. Richard Casdorph, M.D., Ph.D., of Long Beach, California, Assistant Clinical Professor of Medicine at the University of California at Irvine, independently utilized radioisotope studies done at Long Beach Memorial Hospital that indicated how sixteen out of eighteen of his patients with angina showed an increase in the ejection fractions (force of blood flowing through the heart) on radioisotope studies, as well as clinical improvement from chelation treatment.[35] Furthermore, Dr. Casdorph studied a second patient group and found that fourteen out of fifteen patients having cerebrovascular disease with senility developed better brain blood flow and remarkable improvement in brain function. Dr. Casdorph suspects that one mechanism of action of EDTA in these patients is "slowing calcium currents" which are necessary for muscle contraction in the arterial wall, resulting in dilation of the arteries.[36]

9) Effective chelation therapy is not yet available in the United States in tablet form, although some of the new calcium channel blockers recently released for sale in the United States have oral forms available. FDA regulations have kept such excellent oral antiarteriosclerotic drugs—Anginen from Japan, Lisater from Mexico, and Syntrival from Germany—away from the American people. These products work partially through chelation and can be taken by mouth. If you contact the Association for Cardiovascular Therapies, Inc. you can learn more about these oral chelators from overseas. They work remarkably well to unblock clogged arteries.

Most physicians are more likely to prescribe a drug which can be administered by mouth rather than the more time-consuming EDTA intravenous injections. Many physicians will therefore be very quick to accept and prescribe the calcium blockers, although they are still relatively new and untested in the United States. They have no long history of human use to establish proof of long-term safety such as we have with intravenous EDTA.

10) Although only approximately 5 percent of EDTA is absorbed from the gastrointestinal tract, some physicians, not wishing to be bothered with the expense and effort of the intravenous route of administration, have utilized oral EDTA in the treatment of chronic lead or other heavy metal poisoning. This oral use of EDTA is considered dangerous by toxicologists and poor medical practice by the American Academy of Medical Preventics. The oral administration of EDTA tends to increase systemic absorption of lead, since it gets chelated from the gastrointestinal tract into body cells. These negative reports regarding the use of EDTA by mouth in lead poisoning have further confused the average orthodox physician about the potential value of chelation therapy.

11) A particularly devasting negative conclusion in one of the eight articles on chelation therapy written by Drs. Kitchell and Meltzer—their 1963 "reappraisal" of the treatment—has been quoted repeatedly by opponents of chelation therapy such as the National Research Council as the definitive statement on the subject.[37] If one carefully reads the other seven favorable articles by Kitchell and Meltzer, and then focuses on the two paragraphs immediately preceding this eighth widely quoted "negative" conclusion, you will find that the authors had found chelation therapy for the legs actually was useful. The two researchers reported that chelation caused "improved blood flow" in small arteries below the knee, but not in larger vessels. Their statement clearly indicates that they did find benefits from chelation therapy in vascular disease even though Kitchell and Meltzer were incorrectly using it as a "magic bullet."

When all of their eight articles are reviewed, it is clear that Kitchell and Meltzer's apparently negative conclusion requires reconsideration in the light of new knowledge. A knowledgeable reader who studies their work and has passed the AAMP qualifications for diplomate in chelation therapy will soon see that the two doctors did not accept the concept of calcium involvement in hardening of the arteries the way it is understood today. Also, they clearly failed to replace depleted essential trace elements now recognized as involved in arteriosclerosis and heart disease. Finally, they ignored all the obvious risk factors in their patients such as smoking and incorrect diet.

12) Excessively large doses of intravenous EDTA were originally injected much too quickly into patients in the 1950s by the early medical researchers. The result was that some people died not too long after the treatment and EDTA was blamed. These are cases which I shall report on in Chapter 6. As much as 10 grams of EDTA injected daily within ten to fifteen minutes were utilized, and later changes seen in the kidneys were con-

sidered evidence of the "apparent nephrotoxicity" of the drug. Without realizing it, a classic mistake in medicine had been made.

It was later found that so-called chelation deaths were actually due to the toxic effects of lead on the kidneys. Too frequent and too fast infusion of high doses of EDTA will send excess lead from the blood suddenly into the kidneys. Lead and other heavy metals are well known to be toxic to the kidneys.

Such incorrect conclusions nonetheless produced a distorted picture regarding the kidney toxicity of EDTA, which still persists with the careful assistance of the opponents to chelation in organized medicine. The dosage schedule as currently employed by doctors following the American Academy of Medical Preventics Protocol makes EDTA safe. Most orthodox medical doctors remain misinformed by their own medical journals about the safety of EDTA. They will not take the time to read the excellent studies regarding the nephrotoxicity of the ligand done by pharmacologists Drs. Doolan and Schwartz.[38]

13) Clinicians not fully informed about the way EDTA acts in the human body have expressed fear that patients might lose essential calcium from bones and teeth in addition to the pathologic calcium in the arteries. In fact, one research study on rhesus monkeys that were administered repeated EDTA infusions did show loss of skeletal calcium. However, the study's procedure was faulty. The monkeys never received mineral supplements and were confined to small cages. They weren't ever allowed to exercise. Lack of exercise alone will promote loss of calcium from the bones. This faulty protocol, often cited by nutritionist Nathan Pritikin in opposition to chelation, has also hampered the cause of chelation therapy.

14) Medical and health insurance companies have generally refused reimbursement for claims from patients who had received chelation therapy and its related testing. This failure to reimburse policy holders stems primarily from a special meeting held by the Social Security Administration a dozen years ago to formulate an official policy regarding whether Medicare should pay the costs of chelation therapy. Immediately following this meeting, a directive to all Social Security offices was received forbidding Medicare reimbursement for chelation therapy when administered for hardening of the arteries. Yet, no testimony was permitted to be taken from chelating physicians despite their flying to so testify at the Washington, D.C., meeting called by Theodore Bidwell, then the Director of the Social Security Administration.

After Medicare stopped paying for chelation therapy, the other major private health insurance carriers adopted the same policy. Although reimbursement does occur on a sporadic basis, most health insurance companies deny the claims on the basis that EDTA chelation therapy is not "usual, reasonable or customary" when employed for the treatment of hardening of the arteries. This has caused patients to subject themselves to open heart surgery and other procedures generally reimbursable by health insurance policies.

The Medicare rules are now changing. The whole of Chapter 10 is devoted to Medicare and private health insurance coverage for chelation therapy.

15) Before the Kefauver-Harrison Act of 1962 was passed to protect the public from harmful and ineffective pharmaceuticals, following the thalidomide tragedy, the package insert for EDTA listed the drug as "possibly effective" for hardening of the arteries. It was listed as useful in the treatment of angina, vascular claudication, and cerebrovascular disease. After 1962, the FDA was required by law to demand that Abbott Laboratories and other pharmaceutical manufacturers of EDTA prove beyond a reasonable doubt that the drug was effective for the indications contained in the package insert. But the patents on EDTA were expiring and the required medical research to substantiate prior claims would cost the manufacturers from twenty to fifty million dollars and take from eight to ten years. They were forced to remove the "possibly effective" claim by failing to make the heavy research investment, leaving behind with no explanation the misleading statement that "it is not recommended for the treatment of generalized arteriosclerosis associated with advancing age." This statement was probably allowed to stay as a legal protection from potential product liability actions against the manufacturer.

Today, the FDA will not remove the misleading warning without the manufacturer initiating such a request and providing substantial documentation to justify any change. With EDTA no longer being a patentable drug by itself, no pharmaceutical company will make the investment for such research documentation. And the misleading statement continues to appear on the package insert, frightening away any orthodox physician who might consider using EDTA for hardening of the arteries.

16) Historically, the dead hand of orthodoxy has routinely delayed the advance of knowledge in medicine. Most major medical developments such as vitamin C for scurvy, antisepsis, niacin for pellagra, and other more recent modalities have required over thirty years before gaining general acceptance. The status of EDTA chelation therapy is no different from those other medical advances. Physicians have a well-known resistance to change. It is an attitude instilled as a part of their medical training and philosophy. One reason is the burdens they carry with making decisions on life and death. But also the orthodox doctor's tendency is to avoid colleague criticism and stick to mainstream medical concepts in the treatment of patients. A doctor who chooses to evaluate some innovative approach on an informed and willing patient may see himself labeled a "quack" by his

fellow professionals. Pending research grants, promotions, staff appointments, and referrals from other doctors have been known to dry up and disappear from his stepping out of the medical mainstream. Social and professional ostracism discourages most physicians from using procedures considered too innovative.

17) Most orthodox physicians are trained to only accept the results of double-blind studies. Performing double-blind tests with EDTA chelation therapy can't be done, because patients are aware of the drug's action in their bodies from various symptoms I shall discuss in the next chapter. No one is likely to accept a placebo when he is enrolled in a treatment series for reversing or preventing hardening of the arteries. It requires sitting for four hours each time over twenty or thirty treatments. While placebo testing is standard procedure for trials with new drugs on terminal patients, chelation physicians don't consider it justified to give a placebo to someone dying from vascular disease who could be saved by giving him the real EDTA. Additionally, a doctor's faith in the therapy is an integral part of the healing process.

18) Before the establishment of the American Academy of Medical Preventics six years ago, there was no group of physicians competent to judge the standards for delivering chelation therapy. The AAMP went about setting those standards and in October 1980 adopted a protocol for the procedure. But until now, nonchelating orthodox doctors were sitting in judgment on physicians using a therapy with which mainstream medicine was unacquainted.

19) Practically all physicians have been inadequately trained in the new field of using minerals for an organic effect—bioinorganic chemistry—of which chelation is the basic biological mechanism. Indeed, the entire pharmacology and therapeutic applications of chelating agents in medicine are totally new. The average doctor simply does not understand how and why chelating agents work, even in simple lead poisoning.

20) It's not uncommon for physicians to fail to read their own medical literature in which chelation therapy is extensively described, discussed, and explained. While it's difficult for a medical scientist to get a positive chelation therapy article published in the more prestigious medical journals, over 1800 medical journal articles can be found in the medical library with a careful literature search. Approximately 20 percent of the available chelation treatment bibliography is listed in the appendix of this book. It's unlikely that any doctor who ignores published information on chelation therapy will employ, recommend, or even consider such a treatment in any positive way. He will be more inclined to discourage his patient away from the procedure, and this has been the pattern usually followed in the past.

CHAPTER SIX

The Costs and Safety of Chelation Therapy

As with any therapeutic modality—especially a controversial one—every chelation therapy patient faces the same predicament. Given the present negative climate in which this treatment is administered, answers must be known to certain important questions: How much is this procedure going to cost me? How much of my productive time will it take? What are the side effects that go with the injections? Does it hurt? Will chelation therapy interfere with other medicines I am presently taking? What am I letting myself in for?

A lot of people ill with cardiovascular problems have been forced into evaluating their risk-benefit ratios. They needed to weigh the cost effectiveness of their investments of time and cash for getting cardiovascular repair. It all came down to their deciding whether to believe the criticism of the procedure and its medical proponents in the United States or to accept as truth the testimony of 300,000 Americans and another estimated eighteen million patients from around the world who have already undergone chelation therapy.

There definitely are costs! Undoubtedly, the chelation answer to hardening of the arteries takes its toll in money, time, and effort to learn if this treatment seems appropriate for you. Additionally, you must confront its controversial nature, slowness of action, possible toxicity, and frequently uninsured financial cost. You may also feel some definite anxiety about the treatment's potential side effects because the proponents, in order to legally protect themselves, make you sign an informed consent to your acceptance of the potential of death—no matter how remote. To invalidate the procedure, you're quite aware that the critics of chelation are anxious to testify against a chelation doctor if anything ever did go wrong. Other chelation patients and the doctors administering to them are likely to have told you that minor discomforts are rather commonplace, such as a local stinging sensation where the fluid enters your arm, or a sore backside from sitting for four hours waiting for the fluid to run in.

A retired foreman on permanent disability from the United States Steel Corporation, George Kavic of West Mifflin, Pennsylvania, faced this sort of trying situation. Kavic, who lived on a pension, did not have the "cash on the barrelhead," as he describes it, required for extensive chelation therapy. Nor did he have the opportunity to go earn the extra money by taking odd jobs, inasmuch as the man was told by his doctors, "George, don't make any definite plans. You have only three to six months to live!"

Kavic suffered constant chest pains and burning sensations because three-fourths of his myocardium had been damaged by a series of heart attacks. The angina he felt was so great it took him five minutes to move just ten yards across a room. Additionally, his right leg circulation was so impaired, it turned blue and claudicated with any movement or attempts at weight bearing. Kavic's doctors said that they hoped the leg would not have to be amputated, since the patient's heart and general health were so poor he would never survive the operation.

At our interview, George Kavic explained, "In November 1975 my cardiologist told me the results of my angiogram. He said that bypass surgery wasn't possible even though I needed it in both my leg and my chest. I had blockage of two coronary arteries that narrowed their openings by 80 percent and 90 percent. The risk that I would die on the operating table was too great.

"I asked the cardiologist what he thought about my taking chelation therapy, and he said that I was nuts to consider it! He said also, 'If you are still living in September [1976] then make an appointment with me for that following December. You may make it that long, but you have a very tough case.' So I was left waiting to die, but while waiting I had terrible and unbearable pain in my right leg," Kavic said. "I didn't tell my wife because it would have frightened her to death. My leg was blue up to the ankle, and I couldn't walk at all."

Out of desperation and over his cardiologist's objections, the patient telephoned one of the chelating physicians listed in Appendix I to inquire about the chelation treatment. Learning of the therapy's cost, Kavic knew that such funds were out of his grasp. But he knew, too, that no alternative lay open, whereas chelation therapy did offer some possibility of relieving his condition at least a little. He hoped he could squeeze another year of life out of having the treatment. Therefore, George Kavic borrowed the money from loving friends and relatives to undergo chelation treatments.

The long trip to receive the intravenous drip technique with EDTA had Kavic almost exhausted. He experienced feelings of tremendous anxiety. "I was scared," Kavic confessed. "I wondered if this trip was worth it."

The patient took thirty-two EDTA intravenous infusions at that time. Physical improvement had begun for him at about the fourth injection. He telephoned home to report that he was able to walk a mile without feeling chest pains or leg pains. After five weeks Kavic's pedometer showed that he walked six miles a day without any difficulty.

George Kavic told me, "I haven't had the return of symptoms. I don't take any medication, except with each meal I add five food supplements that I buy at a health food store. I swallow these vitamins with my food, and I feel like a million dollars. I walk regularly four, five, and six miles a day because that's part of my rehabilitation program—regular exercise. I returned for a recheck of my laboratory tests about ten months after the chelation treatment. The readouts were magnificent. My triglycerides, cholesterol, and other tests all showed improvement over the first time. Then I had five more bottles of the EDTA solution too, for a total of thirty-seven treatments."

At this writing George Kavic looks in excellent health. He is energetic and forceful—so much so that occasionally he pushes himself to the point of heart pain. Those pains remind him that once he was told he had less than six months to live.

THE VARIOUS COSTS OF CHELATION THERAPY

The financial investment required to finance chelation therapy varies around the country from a low of $48 per treatment to a high of $100. On the West Coast of the United States it tends toward the higher range. The fee along the Eastern Seaboard averages about $85 per treatment. Some of these "price" differences relate not so much to the actual chelation infusion as to the adjunctive measures that are employed as well. Each treatment, for instance, may require you to take megadoses of vitamins and minerals. They sometimes are injected because your body is seriously deficient in these nutrients. Furthermore, some areas of the country have lower overhead costs for the doctor to pay, which all greatly affect these prices.

The chelating physicians invariably request hair, urine, blood, and/or saliva and feces tests in order to determine potential vitamin and mineral deficiencies. The laboratory costs mount up to another few hundred dollars. They also do a diet analysis, so that you may receive the orthomolecular nutrition program specific for your need. Diet analysis forms you fill in are among the most lengthy but thorough investigatory questionnaires most patients will have ever seen. Results are returned to you from a computer. The computer printout tells you what food deficiencies or imbalances you have and what you can do to improve your nutritional life style, and why. It provides the latest research information.

Most of the physicians whom I have met who use chelation therapy and the other holistic approaches in their practice of medicine routinely apply the patient information from these questionnaires. The doctor and you, his patient, are able to help identify problems that are likely to aggravate your health problems and what you can do to sidestep those problems.

Twenty to 30 chelation treatments is a usual complete, initial series producing results. However, 20 treatments are considered the *minimum course* for altering artery pathology that is *mild*, or in its early stages. If given primarily for medical *prevention* rather than as a corrective therapy, the number of infusions administered in a series may be less. Certainly 20 treatments is not enough to *reverse* a condition of hardening of the arteries that is *severe* (50 and even 100 treatments have been required), but the series of just 20 treatments can produce markedly improved functional changes throughout the circulatory system. For the more severe occlusive arterial problems, probably 30 or 40 treatments over a fifteen-week period will be recommended for a single series of chelations. With various tests and examinations, the initial total cost is likely to be $2000 or more.

Among the larger problems faced by people who want EDTA chelation for preventive or therapeutic purposes is the expenditure of time. Financial cost alone is not the whole obstacle. An individual will ask himself, "How can I visit the doctor's office twice a week and remain there for three or four hours each visit? Can I afford the time? Won't I lose money? Who will run my business? How can I explain it to my boss?"

Most physicians find it difficult to get people to invest this much time and money in themselves. The motivation must come from within the individual. A person has to reorder his priorities if optimal health and prolonged life are part of one's major desires.

On the other hand, heart pain that strikes an individual cannot be argued with. Days, weeks, and months can always be found for bed confinement to recuperate from a heart attack. If circulatory distress has turned a leg gangrenous, somehow a person finds the time to have the amputation that will save his life.

"I was no different," said Warren M. Levin, M.D., of New York City. "I decided I was going to take chelation several years ago, and I just never got around to it because I thought I couldn't afford the time. Finally I told myself that I've just got to do it! I managed to arrange for the therapy at the end of my full working day. I ate dinner, relaxed a little, then got into bed, put the needle into my vein, and went to sleep. A registered nurse, whom I had hired at regular nursing rates, sat by my bedside all night for a full eight-hour shift to keep the intravenous drip going at the correct speed. She made sure the chelate did not infiltrate into surrounding tissues. The financial cost to me actually was greater than it is to one of my patients. I was paying a nurse's wages for all those hours that I slept."

Individual treatment times vary for patients, but on the average EDTA administration takes four hours. It will be shorter—as little as two and a half hours—if the patient can tolerate a faster drip. It could be longer—as long as eight hours—if the patient has too much discomfort or any other sensitivity reaction to the components in the EDTA solution.

From one to three treatments per week is a usual number, two per week being quite common, depending on how ill the patient is and what time he has available. This will be the schedule except in those instances where the patient has traveled far for a short stay in a hospital facility or doctor's clinic. Contrary to the conservative routine prescribed by the American Academy of Medical Preventics Protocol, when hospital confinement is required treatment may be given for five days in a row with only a two-day pause to rest the kidneys. Then another five days of consecutive injections are given again. Inasmuch as the EDTA has insufficient time to leave the body, consecutive treatment is slightly more risky. The five-day treatment schedule could be followed even out of the hospital when the patient is in greater cardiovascular danger and more aggressive measures are required. For over twenty-five years, this five-days-in-a-row schedule had been safely applied to thousands of people when they needed chelation therapy. In 1980, however, the AAMP did a reevaluation and discarded the five-day treatment schedule. Consecutive days of intravenous EDTA are not routinely recommended currently. The new diagnostic methods also allow patients' needs for chelation to be recognized much earlier than previously so that a more leisurely schedule can be followed. Moreover, chelating physicians have to be more cautious today because of current medical malpractice conditions.

Many chelators now opt for the conservative program of one, two, or maybe three treatments per week, as well as a lower dose of EDTA based on your weight and kidney status. As more doctors begin offering chelation therapy in more cities. It will be possible to avoid the expense of lodging, travel, and meals altogether. The treatment will become less disruptive to patients' busy work schedules.

Chelation administration technique does not take much skill on the part of medical professionals. Amost anyone can give it. Even paraprofessionals perform the technique well. That is one of the aspects of this treatment that really upsets super sub-specialists in organized medicine who have spent years learning to do vascular surgery; minimal extra training is required. Any physician anywhere could give it to his patient. The vascular disease treatment monopoly is taken from the hands of the specialists such as cardiologists, internists, vascular surgeons, chest surgeons, anesthesiologists, and various maintainers of sophisticated hospital facilities. The rendering of medical services is handed back to the family doctor when he or she chooses to administer chelation therapy.

The generally trained medical practitioner or his nurse can take on the responsibility of the administration of the intravenous feedings, and that is the key here—*responsibility*. A medical professional has to be able to responsibly react to the patient's possible discomfort during the administration of the intravenous solution or to any side effects afterward. This simply means that the doctor should be knowledgeable in the technique. He or she must attend medical meetings oriented to chelation therapy. He or she also must tour the facilities of other chelating physicians and spend time apprenticing at chelation colleagues' clinics. Weekend workshops have to be attended, too, as part of the American Academy of Medical Preventics requirements. Of course, the doctor who chelates has to do extensive reading on the subject, and should subject himself to the written and oral exams that lead to becoming a diplomate in chelation therapy.

Then, you might wonder why, if the treatment is so easy and so simple to administer, the cash costs should be relatively high. There are several pertinent reasons to explain the charges.

First, you must be in the medical office for so many hours. During that time the physician and his staff are entirely responsible legally for your well-being. Patients pay for the physician to take that time-consuming responsibility.

Second, chelation physicians usually require larger offices and more employees to accommodate the huge amount of traffic that moves in and out. This results in significantly larger overhead costs to pay.

Next, the physician who chelates must take the time from his practice for attendance at the frequent and expensive medical conventions required for him to remain abreast of new developments in his field and to learn more about bioinorganic chemistry. It is an ever-changing field.

Finally, the chelating physician usually has to retain several competent attorneys. Patients, you and I, have to help pay for the doctor's additional legal risk and exposure due to the unrelenting peer pressure he gets from organized medicine when he offers chelation therapy to his patients. Chapter 11 tells of the politics involved in chelation therapy. You will read of the intra- and interprofessional stress physicians must face for daring to offer EDTA chelation to their patients.

THE INTRAVENOUS INFUSIONS YOU MAY EXPERIENCE

Your experience with receiving chelation therapy is pleasant, although monotonous, for the first few three- to four-hour procedural visits, until you learn to arrive prepared to accomplish some reading or writing during your treatments.

Each visit, you sit in a reclining chair and follow a procedure invariably the same for all of the treatments. The physician or nurse acts as "intravenous technician" and attaches a disposable infusion set to the prepared bottle of EDTA solution. An extension tube set and a small vein needle are added. The technician fills the tubing with fluid to expel all the air present. A strip of white adhesive tape is affixed alongside the bottle's graduated markings. When the infusion begins the technician writes down the time the bottle should empty to each successive 100 cc mark. Each bottle in your series of infusions is numbered in that way to be sure it doesn't go in too fast and thus possibly give you less benefit.

The doctor or the nurse, as technician, has the option to select a venipuncture site at your ankle, next to one of the leg bones, or along one of your forearms. A site is chosen alongside and preferably above a firm bone surface away from movable joints and usually not in the hands. This selection thus gives the technician an option to use or not use an infusion board. Your physician or his nurse injects the needle into your limb as part of that office's preferred chelation procedure.

Needle insertion consists of the technician engorging your vein site carefully by massaging and gently slapping the part. Then a small Velcro tourniquet is applied. He or she cleans the skin with alcohol and verifies the fullness of your vein. An assistant to the technician holds the bottle hanging down at a level well below the vein and keeps the bottle valve closed.

Then the technician folds the needle wings together and holds them between thumb and forefinger, the opening upward and pointing in the direction of the vein's blood flow. The skin and vein wall are punctured in one quick motion that is practically painless. The assistant opens the bottle valve to create suction in the tubing. As blood shows in the tubing, the technician releases the tourniquet, flattens the needle wings, and affixes them with a short strip of transparent micropore adhesive tape.

The technician's assistant raises the bottle very slowly while adjusting the drip valve to about twenty drops per minute at the start. The technician affixes successive coils of tubing to the skin with separate strips of tape. A last piece of tape, long enough to go around, is dropped over the connecting glass segment of the needle set and is draped gently around the limb without pressure. Hanging the bottle in its hook on a stand for that purpose, the technician writes down the starting time for the infusion on the bottle's white tape at zero level. Setting up the apparatus takes about as long as it took you to read about it.

Patients are observed for a few minutes each time to assure the technician that the infusion is well tolerated. From time to time during your infusion the drip rate is adjusted to have it run at slightly faster than one hour for each 100 cc. Five hundred cubic centimeters of solution are administered. The procedure averages four hours a visit, while slower rates or even lower doses of EDTA may be used for sensitive patients.

You will experience few and relatively minor discomforts during the infusions. There might be slight bleeding into the tissue, with some black and blue swellings.

Since the material being infused is somewhat irritating and different people have various levels of sensitivity, it is to be expected that you could feel discomfort from the solution going through your arm. This is not usual. An anesthetic can be safely added to the solution to prevent most discomfort if you find it to be a problem.

IS EDTA SAFE FOR THE TREATMENT OF THE ARTERIES?

In laboratory investigations on rats and mice, comparing the toxicity of EDTA injection therapy with other drugs commonly used by Americans, it was found that an equivalent therapeutic dosage of EDTA is safer than one aspirin tablet, a dose of digitoxin, a tetracycline capsule, a teaspoonful of ethyl alcohol, or the nicotine in two cigarettes.[1,2,3,4]

Citizens of most of the countries of Europe and all of the United States hardly let a day go by without using EDTA products on or in their bodies. It is made a part of literally thousands of common preparations such as foods, flavorings, soft drinks, animal feeds, plant nutrients, herbicides, pharmaceuticals, biologicals, fungicides, germicides, cosmetics, creams, oils, soaps, ointments, bath preparations, hair dyes, hair wave products, metal cleaning solutions, pulp and paper, scale removers, rubber-coated fabrics, textile preparations, rubber polymerization, water-softening agents, emulsion stabilizers, leather processors, photographic agents, radioactive decontamination, organic systems, and more.[1] Fifty food companies around the globe put it into their canned goods, packaged products, container liquids, bottles of fruit drinks, baby food, and other items for human consumption. EDTA is an integral part of our lives, so that it had better be safe.

In the field of toxicology, the poisonous quality of a substance is determined by its LD-50. The LD-50 measurement is based on the toxic ingredient being able to produce death for 50 percent of the animals being tested. In actuality, you must have two particular values to discuss the toxicity of any drug. One value is the therapeutic or optimum effective dose and the other is the toxic dose of LD-50. *This LD-50 is the dose that, over a specified period of time, will kill half the animal subjects to which the drug is being administered.* Then the toxic dose divided by the therapeutic dose is the *therapeutic index.* This therapeutic index tells researchers and physicians how dangerous a drug or other substance is when administered to living organisms, especially people.

If the therapeutic index is small, the compound or drug is toxic; if it is large, it is nontoxic. For example, if 4 gm per day is the optimum dose of a drug and 8 gm per day is the LD-50, the therapeutic index is two. Obviously this would be a very dangerous drug because the therapeutic index is small. Insulin may fall into this dangerous class. It has a low therapeutic index, and users know that insulin has to be administered with great caution. EDTA has an exceedingly high therapeutic index and is safer dose by dose even than aspirin.

Illustrating this in practical terms, the LD-50 of EDTA is approximately 2000 mg per kg of a person's body weight, based on oral repetitive studies. (The same is true for intravenous administration.) And these studies have been extensive and exhaustive over time. The toxicity of one infusion of EDTA can be compared to something we already are familiar with—aspirin, the most common drug taken for pain. Aspirin's toxicity is 558 mg per kg of body weight. So essentially, EDTA is about three and a half times safer or *less* toxic than aspirin.

Aspirin is legally used as the treatment of headache, but EDTA has a much more important use. It improves blood flow to the entire body, including the heart, the head, and the limbs. The chelating substance brings about this improvement partly because one gram of it is able to help sequester at least 100 mg of calcium from your body, some of which comes from the calcified cells in arterial walls. In some really seriously ill patients, well over 600 mg of calcium have been recovered in the twenty-four-hour urinary collection following intravenous infusion with the maximum dose of 3 gm of EDTA. The exact amount of calcium excreted by each patient, however, depends on many variables, including the status of his or her parathyroid glands, the amount of calcium and phosphorous consumed in the diet, and the ratio of these two minerals in the body. It also is determined by whether the patient is "fortified" with too much vitamin D_2 eaten in his cereal grains, milk, and other foods, his magnesium and protein intakes, and certain other factors.

In any case, serum ionizable calcium levels are immediately reduced to 50 or 60 percent of pre-treatment levels with intravenous injections of EDTA. If the physician doing the injecting closely follows the Protocol established by the American Academy of Medical Preventics, you should experience no significant toxicity. If administration is improper, such as an excessive EDTA dose or possibly a too-rapid rate of infusion, certain side effects or even potentially toxic problems could make their appearance. Risks such as spasm and twitching of the face, hands, and feet from too much calcium being lost too quickly, convulsions, severe cardiac arrhythmias, and even respiratory arrest theoretically could occur. Careful administration technique as described in the AAMP Protocol, which anyone can obtain by contacting the Academy, is the key to avoiding these risks.

The trained chelator gives a slow infusion of dilute solutions of EDTA. Dosages never exceed 3 gm or 50 mg per kg of body weight, so that a sixty-six-pound person would only receive 1.5 gm as the total dose. Lesser doses are given to children, the elderly, lightweight adults, and patients having their initial infusions.

Extra precautions adopted by the physician specializing in administering EDTA allow adequate mobilization of calcium ions from the calcium stores of the body. This is accomplished by the parathyroid gland, which is responsible for making calcium adjustments. When EDTA injections pull metastatic calcium out of the places where it doesn't belong, a roundabout chemical and physiological mechanism is set into motion. The chelation sequence takes place first among calcium ions floating in the bloodstream. The metastatic calcium is then forced by the parathyroids to replace the mineral missing from the bloodstream, ionic calcium. Giving parathormone, the hormone secreted by the parathyroid gland, this chance to function appears to be an essential aspect in mobilizing the pathologic or metastatic calcium stores from the walls of the blood vessels.

The earliest medical researchers using EDTA chelation therapy for reversing hardening of the arteries did not have the present knowledge. Lack of knowledge contributed to their confusion about how chelation therapy works, and why the maximum benefits usually appear three months after you complete a series of these infusions. It also caused some side effects which have haunted chelation therapists into 1982.

NEPHROTOXICITY OF CHELATION THERAPY, IF ANY

The initial use of EDTA in 1954 was associated with two deaths[5,6] for which serious overdosages of up to 10 gm per EDTA infusion are blamed. Later, the recommended dose was lowered to 5 gm and then to 3 gm;[7] but it was only after the development and use of the electron microscope and other careful studies that the effect of EDTA on the kidney cells was fully clarified.[8,9,10] The conclusion of these studies was that the term "nephrotoxic" or kidney poison is not justified for EDTA. The change seen in the kidney is a normal physiologic mechanism for removal of toxic products through the kidney. There is *no* long-term damage or development of later kidney complications associated with EDTA when the intravenous technique is properly employed.

Norman E. Clarke, M.D., wrote a report on the death of one patient who received EDTA. This occurred during his early investigation of EDTA. One of Dr. Clarke's chelation cases, a sixty-eight-year-old mechanic identified as W. McL., who had his first attack of angina in 1947, possibly might have had his death hastened from a high-dose-related EDTA infusion. That was almost thirty years ago, and there has been no documented case like it since then. Nevertheless, the deaths have left EDTA chelation therapy with an undeserved reputation for being dangerous and nephrotoxic. The old incidents frighten many doctors away from using it to treat their patients.

Dr. Clarke stated, "The patient's activities had been unlimited until early 1953. Then he had a diseased gallblader removed in November 1953, after which he improved until January 1954, when angina pectoris increased rapidly."

In February 1954, Dr. Clarke gave the patient fifteen intravenous injections of 5 gm each within less than three weeks (a total of 75 gm). This amount was about five times what is now considered to be the usual and customarily acceptable dosage. A few days later the man had a convulsion, lost consciousness, and died within a few hours. His autopsy disclosed that he also had previously unrecognized kidney disease, which further points out the necessity for first determining kidney function in all patients before they receive chelation. Many forms of kidney problems are commonly not diagnosed until autopsy. The modern chelating physician checks your kidneys routinely now.

Dr. Clarke's patient was probably doomed to die momentarily no matter what treatment he had received.The disease in the man's kidneys alone could have been the cause of his death. Or the mechanic could have died from coronary artery blockage. No actual proof was ever found that would incriminate the EDTA Dr. Clarke had given the patient.

As an explanation of that long-ago death, Dr. Clarke has said, "That was during the early period when we were giving up to 10 grams of EDTA. We didn't know what was the optimal dose to give, it being the first use of the compound on atherosclerosis of humans. We soon found that it was too high a dosage. We reduced it to 5 grams and later decided the correct dose was established at 3 grams."

There has not been a reported proven death from the proper administration of intravenous infusion of EDTA (as described in the AAMP Protocol) for at least twenty-eight years. To emphasize, the dose given today is 3 grams or less of EDTA for a usual number of one, two, or perhaps three treatments per week, each delivered in approximately four hours.

New research on chelation therapy and kidney function was presented by Edward McDonagh, D.O., of Kansas City, Missouri, at the AAMP November 1981 semiannual scientific conference. Dr. McDonagh demonstrated that properly performed chelation therapy actually improves kidney function in most cases. It seems that chelation treatment may be an excellent way to forestall kidney failure in many renal conditions, particularly when the renal disease is due to vascular impairment, which is a not uncommon problem in our Western industrialized society.

Yet, no one should ignore the possibility of kidney involvement. Nephrotoxicity (poisoning of the kidneys due to the removal of heavy metals such as lead) is a serious potential hazard of chelation therapy with EDTA. If signs of nephrotoxicity should develop or be present in any form, heavy metal toxicity—lead poisoning in particular—must be carefully looked for through sophisticated laboratory analysis. Lead poisoning of the kidneys could mistakenly be attributed to the EDTA.

Heavy metals such as lead, aluminum, mercury, cadmium, nickel, copper, and arsenic are nephrotoxic. This is particularly true when the heavy metal is being eliminated through the kidneys, as happens with heavy metals bound to EDTA.

Toxic metals are loosely defined as those elements whose presence at certain low concentrations are known to interfere with normal functions of the cells, usually by damaging the enzyme systems. Excessive levels of toxic metals are exceedingly common in people in industrialized Western countries, where we are all exposed to serious heavy metal pollution. For this reason I shall discuss lead poisoning in Chapter 8.

Nephrotoxicity from heavy metal poisoning induced by EDTA may show symptoms or signs that are usually readily reversible and clear themselves within a few days after the drug has been discontinued. They may include urinary urgency, frequent urination at night, the passage of large quantities of urine, painful voiding, the inability to urinate, swelling or edema, the passage of only a scant amount of urine. There may only be the appearance of protein or blood cells in the urine when it is examined in the laboratory, so such an examination must be done regularly on patients who are receiving chelation therapy. In extremely rare cases the patient could experience kidney failure, kidney insufficiency, or even a large amount of blood in the urine from a previously undetected bladder lesion showing up. Bleeding tendencies resulting from the lesion are increased for several hours after EDTA administration, due to the lowered calcium levels.

Before a trained chelation specialist gives you intravenous EDTA infusions, he or she invariably checks the minerals in your hair as an indicator of your body burden of heavy metals, to help identify any excess potential for developing kidney poisoning. Regular and frequent reevaluations of the kidneys using urinalysis must be carried out during the course of the treatment. These heavy metals are lingering as potential toxins in your own and anyone else's body from what we ingest over a lifetime in the food, water, and air. The chelation injections pull these poisons out of you. What may be bad about the detoxification is that your kidneys could temporarily become overburdened by the metals going through them too fast at high concentrations. When chelation therapy is performed properly, overburdening of the kidneys is unlikely unless something was already wrong with your kidneys. The doctor must make sure, for this reason, to adequately check your kidney status before you begin treatments. He uses special techniques such as having you collect urine for twenty-four hours to measure it for a urine substance called creatine. If anything unusual is found, appropriate kidney X-rays and possibly even a consultation with a kidney specialist is obtained. Many people have weak or sick kidneys and do not know it. These appropriate steps can usually identify kidney-impaired individuals. They need to be watched even more carefully than usual when they undergo chelation therapy.

The old belief among pathologists that the vacuoles (tiny spaces) showing in the kidneys after a quick EDTA infusion was a sign of kidney poisoning has been proven wrong. The vacuoles have now been identified as the normal physiologic process of pinocytosis, a cellular process of actively engulfing liquid. It occurs in the kidneys as a phenomenon in which minute folds develop in the surface of the cell membrane. These invaginations close to form fluid-filled vesicles. The pinocytosis reverses upon allowing the kidney epithelial cells adequate time between infusions to return to their usual state (from twenty-four to forty-eight hours).

The experience of AAMP participating physicians has been that many patients who were the victims of significantly impaired kidneys before receiving chelation therapy have been safely administered intravenous EDTA for hardening of the arteries. In fact, most kidney-impaired people will actually show kidney function improvement from the EDTA treatment, since blood vessel disease may be the cause of their kidney malfunction. Vascular disease is clearly not the only reason for impaired renal function, and other causes would not be expected to show healing after chelation therapy.

POSSIBLE SIDE EFFECTS FROM EDTA INJECTIONS

The American Academy of Medical Preventics maintains a standing policy of requiring every participating physician employing EDTA to report any previously unrecognized complication, reaction, or side effect of the treatment. The Academy's Scientific Advisory Committee reviews these reports even though they merely are an indirect result of EDTA infusion. Consequently, a lengthy list of possible or potential side effects have been gathered. The side effect may have been reported only once during the three million or more EDTA intravenous infusions so far performed in the United States. Still, they are circulated as warnings to the AAMP physicians.

A listing or discussion of possible side effects or adverse reactions from the administration of EDTA chelation therapy which have been reported to the AAMP Scientific Advisory Committee follows:

- *Pain, irritation, or burning at the site of infusion.* This may occur as a result of the needle slipping out of the spot where it's been injected or a needle being too large a gauge for the vein. You may feel a sharp sensation as if you've been stung by an insect. If this happens, the tube must be immediately clamped off to stop the solution from infiltrating under the skin; it must go only into the bloodstream, where it immediately is safely diluted by the blood. EDTA should not be injected into muscles or the skin, for it will combine with metallic ions immediately in the area and produce severe irritation, including the sloughing of skin. Magnesium chloride, magnesium sulfate, or procaine is sometimes added to the intravenous solution by the physician to reduce any local irritation or sensation of discomfort. Since some of us are more sensitive than others and feel uncomfortable even when the solution is correctly going into the vein, these additions ease the injections.

- *Temporary sensations of tingling, numbness, or pins and needles around the mouth.* Neurological involvements of this type come from a too-fast infusion.
- *Fever.* Usually fever occurs if you have a cold or are getting the flu; it is also possible that the treatment itself may induce fever. It could be an allergic type of response to one of the several ingredients contained in the chelation therapy bottle, or a possible stress reaction, or even some possible contaminant in the solution. Any of these factors could cause you to have an elevation of body temperature, which always disappears in a day or two. Aspirin or other antifever medicines get rid of the symptom, unless you were already coming down with a flu syndrome the exact day you took your chelation—then it must run its course. Bouts of fever may be shortened by high doses of vitamin C supplements.
- *Nausea and vomiting.* One out of twenty people may experience nausea, which can be prevented by taking extra high doses of vitamin B_6 (pyridoxine). Pyridoxine is one of the additional nutrients that may be pulled out of the body by chelation therapy.
- *Diarrhea.* Intestinal problems are overcome with standard antidiarrheal medications.
- *Headache.* Eating insufficiently before having the treatment may bring on headache.This can usually be avoided by your bringing a snack to munch on during the three to six hours that it takes for the infusion to be completed. Nuts, seeds, bananas, and other sources of protein work well to prevent headache. Of course, aspirin or other analgesics may be employed, as well, to relieve the pain. Many physicians routinely recommend eating one banana during your first hour of treatment. They report that most of the time the side effect of headache is completely eliminated.
- *Dropping of blood sugar.* Hypoglycemia with its typical symptoms of depression, sleepiness, irritability, and other mood changes has been known to occur from treatment with the disodium edetate. By eating something nourishing during the infusion, especially a fruit such as a banana, you will lessen the incidence of hypoglycemic reactions. While you should eat something before having the four-hour infusion, I recommend that you do not consume high-calcium-containing foods such as dairy products. Rather, eat adequate unrefined complex carbohydrates and avoid most sugars, including overripe bananas. The doctor will check your blood glucose levels from time to time in the event that you display any signs of hypoglycemia.
- *Insulin shock.* EDTA chelation therapy gradually decreases the insulin requirements in most people with known diabetes, which is a benefit.

However, patients using long-acting protamine zinc insulin (the type of insulin diabetics may take), which is chelated to zinc, could find that this long-acting insulin is released too rapidly. An overrelease of insulin may produce a profound hypoglycemia with a resultant insulin shock reaction. Treatment, of course, is to get sugar into the patient and switch to a different type of insulin.

- *Excessive fatigue and a feeling of weakness.* Temporary conditions of weakness and tiredness might come from the presence of an abnormally small concentration of potassium, magnesium, or zinc ions in the body cells. Potassium is one of the minerals pulled out of the blood by chelation therapy and should be replaced in supplement form or by eating high potassium foods such as the banana.
- *Joint pains.* Feeling that you have come down with the flu and are aching in the joints when you really have no infection could possibly come from an "overchelation" syndrome. Chelating physicians observe joint pain in a few patients receiving chelation two to three times a week. With its appearance, the frequency of treatments is decreased to once a week. In addition, the dose of ingredients in the infusion may be lowered or the rate of infusion lessened. The doctor will judge what should be done to eliminate the achiness. If joint pain really signifies the flu, go home and take vitamin C.
- *Skin rash.* An inflammation of the skin taking the form of constant shedding or peeling of the superficial skin layers may occur with EDTA chelation therapy, induced by either a zinc or pyridoxine deficiency. Elimination of the rash is accomplished with replacing the missing nutrient. Such exfolistive dermatitis is not seen in patients who are being treated according to the AAMP Protocol because it calls for vitamin-mineral supplementation on a preventive basis.
- *A drop in blood pressure upon standing upright.* It's very possible that the EDTA will so lower your blood pressure during its infusion that upon your standing quickly to go to the bathroom or upon the infusion's completion, you could almost lose your balance and fall down. Such a circumstance will be prevented merely by rising slowly. If you feel weak or dizzy, sit down and have the attendant check your blood pressure. No intervention with drugs has been found necessary to raise the blood pressure. Patients with hypertension usually find blood pressure comes down to a normal range and frequently it stays there for six months or a year, at which time monthly chelation maintenance treatments may be needed.
- *Thrombophlebitis.* Inflammation of a vein with a blood clot forming within the vein occurs occasionally. Moist heat and the taking of natural antioxidant nutrients such as vitamin E, vitamin C, and selenium do a job of protecting against thrombophlebitis. People having trouble with the blood clot problem will probably have heparin added to their EDTA solution by

the doctor. Reportage of vitamin E causing thrombophlebitis indicates that some vitamin E sold in the United States has no actual vitamin E activity in it. The supplement may only contain rancid oil, which definitely could produce thrombophlebitis. Your orthomolecular physician will help advise you regarding the more reliable brands of vitamin E. It may cost more than the cut-rate that are sold by some pharmacies, health food stores, and mail order houses.

- *Congestive heart failure.* Because of the body's inability to handle the fluid load from the intravenous feeding or the sodium content of the intravenous vehicle carrying the disodium EDTA, a patient already suffering with congestive heart failure (inadequate functioning of the heart as a pump, shown by shortness of breath and swollen ankles) could find that his condition is temporarily aggravated while he is being infused. Also, EDTA is antagonistic to digitalis and decreases its effectiveness, which can further aggravate congestive heart failure if digitalis is being taken. A person possessing such a limited heart reserve must receive his injection with a low-sodium intravenous vehicle, such as dextrose or fructose.
- *An alteration in the rate of metabolism.* Someone underfunctioning from hypothyroidism may find himself speeding up in his everyday activities. With improved circulation of a formerly impaired thyroid, restoration to normal is not uncommon, something the patient may find surprising. In contrast, an untreated or unrecognized trace element deficiency induced by the chelation therapy could bring on a slowed-down metabolic state, too. The chelation doctor should correct this by prescribing trace minerals, including iodine, such as diiodatyrosine.

Other adverse reactions or side effects that have been reported to the AAMP Scientific Advisory Committee are chills, anemia, various skin eruptions, a feeling of unwellness, weariness, thirst, muscle cramps, back pain, abdominal cramps, and lack of appetite. All of these symptoms disappear rapidly when the EDTA dosage is cut or the treatment discontinued.

CONTRAINDICATIONS TO THE CHELATION TREATMENT

EDTA chelation therapy is primarily employed for people who are the victims of hardening of the arteries or heavy metal toxicity. Anybody suffering with well-documented heart problems, coronary blockage, stroke, senility, macular degeneration of the eyes, peripheral vascular problems in the upper or lower limbs, or other vascular occlusive disease makes an ideal candidate to have the treatment. As I mentioned, it's usual for the chelating physician to first check you with base-line diagnostic studies. These studies will document the extent of your trouble and show the progress your health has made when the studies are repeated at the end of the course of treatment. Additionally, diagnostic tests are made to learn if, in fact, you are a candidate for chelation therapy at all. There are particular relative contraindications connected with undergoing this procedure.

Taking clinical considerations into account, the doctor must decide if your need, such as in acute lead poisoning, for example, may override any potential contraindication he may discover. In other words, someone must make the benefit-risk ratio judgment, and it's usually the chelating physician who takes the final responsibility for administering the chelation therapy to you.

Acute and chronic liver disease could be a contraindication. The exact diagnosis should be established by liver scans, biopsy, and other tests. These tests would establish whether the problem was cirrhosis, cancer of the liver, chronic or toxic hepatitis, or another serious liver problem. However, chronic cirrhosis of the liver has been helped with chelation therapy.

Kidney disease may or may not be a contraindication, depending on the degree of impairment. Chelating physicians following the AAMP Protocol divide kidney disease into three categories:

- *Mild renal impairment* takes in a loss of kidney function where the twenty-four-hour creatinine clearance test (a laboratory examination for a chemical found in the blood and excreted in the urine) shows results with less than one-third of age-expected values. If the doctor decides you can take the treatment, he will do careful kidney monitoring tests all through the series of infusions. Additionally, the chelator is probably going to lower the standard doses of EDTA given and administer those infusions more slowly and at less frequent intervals over a long period. He'll require you to increase your food and pure water intake, as well, and possibly use some sauna baths to increase sweating. Sweat can help eliminate toxins, even heavy metals, from the body.
- *Moderate renal impairment* would be the classification for a person having more than one-third loss and less than two-thirds loss of his age-expected creatinine clearance values, or other dynamic renal function tests. Such extra tests would likely include radioisotope studies. Riskier patients will get increased kidney monitoring with blood urea nitrogen (BUN) and routine urine tests before each follow-up chelation treatment. After every third chelation injection, the full creatinine clearance test and a complete urinalysis is usually done.

If you are a moderately impaired kidney patient, you won't take chelation therapy more than once every week or two. You will receive one-half or less of the calculated dose by weight of EDTA, until you've shown sufficient improvement to fit into the "mild renal impairment" category.

- *Advanced renal impairment* of a patient usually means he is not a chelation therapy candidate at all. Such an unfortunate person will have grossly diminished kidney function that shows a two-thirds loss of age-expected values on creatinine clearance testing or equivalent renal function tests. On the other hand, if the patient's hardening of the arteries condition is so advanced that it presents an immediate serious risk of death, the benefit-to-risk ratio may warrant the doctor's initiation of a careful trial of some chelation treatment.

In such a dangerous undertaking, EDTA probably won't be the chelating agent used, at least not in the beginning. There are other agents usually put into the EDTA solution that act as chelators on their own. Such agents (to be discussed in Chapter 7) include the usual carrier vehicle itself, lactated Ringer's solution, ascorbic acid (vitamin C), magnesium, and other nutrients. Lactated ringer's solution, for instance, contains sodium, potassium, and calcium chlorides with sodium lactate in distilled water. When the patient has upgraded his health, the cautious addition of very low dose EDTA may become part of the solution, but given infrequently. The patient will remain under close monitoring by the doctor, with laboratory tests performed before each intravenous injection.

Active tuberculosis is a potential contraindication for EDTA chelation therapy, unless the patient is under absolute control with antituberculous drugs. People with old walled-off or calcified tubercular lesions, where their lungs show clinically healed or chronic calcified tubercles, are usually allowed to receive chelation therapy. The danger in these lesions is that calcification would possibly be stripped from the tubercle, and the bacillus organism could become active again. Therefore, this potential risk has to be weighed against the need and the potential benefit of taking chelation. The new antituberculous drugs, however, greatly diminish the actual risk, and usually careful monitoring is all that is required.

Lesions occupying space in the cranium, including brain tumors, may represent a relative contraindication. Your history of seizure disorders could also be considered a relative contraindication. The hypocalcemia (loss of calcium) induced during the chelation process theoretically could bring on convulsions. In actual practice, this has not been reported, but a chelating physician takes on the burden of considering cranial lesions in deciding if you should receive the treatment. Again, treatment will revolve around the need and potential benefit versus any perceived risk, no matter how remote.

Pregnancy ordinarily is a contraindication unless the potential benefit of saving a woman from dying of cardiovascular disease hangs in the balance. Safety for the fetus has definitely not been established. Obvious danger to the fetus from induced hypocalcemia exists while it is in rapid growth stages. Birth studies in rats indicated that EDTA injections impaired rat reproduction and caused fetal malformations. However, these malformation effects were later prevented by simultaneously giving the pregnant rats supplements of dietary zinc at the time EDTA injections were administered. Thus, the laboratory animals may have been simply manifesting an induced zinc deficiency from undergoing chelation therapy. Nonetheless, the therapy during pregnancy should be avoided until more is known.

Patients demonstrating gross mineral or vitamin deficiency aren't candidates for chelation therapy until their nutritional balances are restored. Such deficient people may have a malabsorption problem for zinc, manganese, iron, calcium, magnesium, and other essential trace minerals. It's possible they need bowel detoxification, with appropriate fiber including high methoxy-pectin (a carbohydrate found in the rinds of citrus fruits and apples), cleansing with enemas or bentonite (a natural laxative), the reestablishment of healthy bowel flora by eating yogurt, a long-term pancrease enzyme (a secretion extracted from the pancreas), betaine (an acid food supplement), or other forms of acid supplementation. Surprisingly, the common problem of relative hypochlorhydria in our population is mistakenly believed to be stomach upset due to excess acid, when, frequently, it derives from a lack of acid.

DRUG INCOMPATIBILITIES AND INTERACTIONS WITH INTRAVENOUS EDTA

If you are now taking digitalis for a failing heart, you can continue with the medication, but be aware that EDTA chelation therapy may temporarily impair the digitoxin effect. Digitoxin is indicated in the treatment of heart failure, tachycardia (rapid heartbeat), and other heart problems to bring cardiac patients' heartbeat back to normal. Inasmuch as the infused amino acid is rapidly excreted from the body, reduction of the drug's effectiveness by EDTA is transient, lasting only four to eight hours.

An indication for EDTA well recognized by the FDA and orthodox medicine is in treating digitalis toxicity. Nonetheless, the chelating physician's biological monitoring of digitalis levels in your body by means of laboratory tests may be necessary, particularly if you develop any sign of congestive heart failure. The doctor will likely discontinue the chelation therapy in congestive heart failure until medical control of your heart problem has been attained.

Propranolol hydrochloride, a drug known commercially as Inderal (Ayerst Laboratories), is a beta-adrenergic (sympathetic nerves) blocking agent that decreases heart rate, cardiac output, and blood pressure. It has been used extensively to control angina pectoris due to coronary atherosclerosis. Unfortunately, some evidence suggests that the beta-blocking type of drugs may also partially block parathormone (the hormone of the parathyroid) response, so that to an extent, EDTA and Inderal are potentially slightly antagonistic to each other. There are times, however, when both drugs may have to be used, at least until enough chelation has been given to see some improvement. Then Inderal may slowly be withdrawn. You want your body's parathormone level increase to help pull metastatic calcium out of the cells of the arterial walls as replacement blood calcium when EDTA leaves the bloodstream, going out in the urine and carrying away its ionic calcium. Theoretically, with parathormone being partially blocked from replacing the ionic calcium with metastic calcium, the consequent hypocalcemia could take longer to correct. The parathormone-induced osteoblastic activation (to make bones take up more calcium) may be diminished. For this reason, when you can get along without Inderal satisfactorily because the EDTA chelation therapy has begun to work, you may be carefully taken off it.

If you are prescribed other medications such as coronary vasodilators, nitroglycerine, antihypertensives, cerebral vasodilators, peripheral vasodilators, aspirin, or combinations of such drugs, you generally continue with them as needed throughout the chelation treatment program. Frequently the need for these extra medications begins to diminish after five to ten chelation treatments. No impairment of the EDTA therapeutic response in the body seems present when chelation therapy is administered with such drugs. Be cautioned, however, that coffee drinking can cause an excessive release of prostaglandins, possibly by stimulating production of prostaglandin synthetase. This extra prostaglandin may have an unfavorable effect for patients on intravenous EDTA. Therefore, avoid coffee drinking.

Many patients will find that they improve sufficiently on the complete EDTA chelation therapy program to get along with far fewer drugs. Some drop all the pharmaceuticals they once thought they couldn't live without. Yet, chelation patients may now take even more pills. They swallow food supplements, which are far safer.

Cigarette smoking does block much of the therapeutic response to the treatment. Smokers must make every effort to discontinue the inhaling of nicotine, cadmium, carbon monoxide, tars, and other poisons present in cigarette smoke. Nicotine is especially bad for a person with vascular disease, since this drug induces vasospasm. You are already trying to cope with arterial cells that have excessive levels of intracellular calcium in them. This injurious calcium activates the contractile apparatus in arteries. Once activated, the affected blood vessel may stay in vasospasm for hours, so that even a few cigarettes a day will still be enough to seriously impair the benefits from chelation therapy. Furthermore, the nicotine in cigarettes blocks your body's production of the critical prostaglandin PGE. PGE is involved in keeping blood vessels dilated. When it is lacking or deficient, blood vessels automatically respond to its ever-present antagonist, and go into spasm.

After completing a full course of the prescribed number of treatments, you should return to the chelating physician once a year for reevaluation. The same testing he had done for the treatment's earlier short-term evaluations must be repeated in order to compare your current status with what it had become at the end of the treatment series. Try to go through this health reevaluation every year for five years. Documenting your cardiovascular wellness will tell you if you need any additional intravenous injections with EDTA.

After you've been feeling well for a long time, you *do* tend to forget how ill you once were. It's not uncommon for previously chelated people to let down on their improved life-style and resume the unhealthy way of living that brought them to needing chelation therapy in the first place. Nobody wants the disabilities of hardening of the arteries to start all over again, and annual noninvasive retesting of your circulation keeps you honest. It also informs your doctor of any recurrence of your impaired circulation. Before damage occurs, he can take corrective action.

You can help avert the rapidly spiraling costs of hospital care. Avoid getting a heart attack or stroke. Let the hospitals empty or let them specialize in accidents, or better yet, let them convert over to preventive medicine, including chelation therapy. You can help force medical care into this transition by not getting vascular disease that is responsible for so much of hospital use today.

CHAPTER SEVEN

How to Chelate Yourself at Home Using Common Substances

Could chelators be swallowed as food or a pill? Without question! Food, drug, and exercise chelators literally do already exist. Natural chelating agents would be the best method—nutrients we could consume as protectors. Certain of our activities, foods, and nutritional supplements help to guard us against heart disease, senility, arteriosclerosis, heavy metal toxicity, and other chronic degenerative disease.

Having spent his early youth in small-town Georgia eating fried chicken, gravy, sausage, eggs, biscuits, and country ham, and with both his father and mother having died of heart trouble, Rollin M. Kimball of Fort Lauderdale, Florida, had stepped into adulthood with a predisposition for heart disease. He had several genetic and lifestyle heart attack risk factors. To them he added heavy smoking [three packs a day], ten pounds of excess weight, and a totally sedentary way of living. He gave no thought to regular exercise except for occasional tennis or golf.

His first angina attack from blockage or spasm in his coronary arteries hit him at the age of forty-nine in December 1974. As a real estate broker, Kimball decided to remain deskbound and "restout" the sensation of throbbing pain in the chest. He left off the walking exercise that he blamed for bringing on the pain.

In June 1975, Kimball woke up one night sensing that he was unable to breathe. His chest felt as if a large weight were resting on it. "I couldn't fully inflate my lungs. An iron hand was around me," he said in a *New York* magazine article.[1] To relax himself he opened the window, tried to breathe deeply, and lit a cigarette.

He visited a group of cardiologists. A stress test on a treadmill revealed that Kimball could exercise only 3.2 minutes before he had to stop because of angina pain. One of the cardiologists recommended a triple coronary artery bypass operation right away. A section of vein would be stripped out of his leg and spliced in, one end before and one end after the blockage in each of the arteries feeding blood to his heart. These three splicings would bypass the narrowed areas in his heart blood vessels.

Kimball turned down the idea. He explained, "I visualized my arteries (about the size of a soda straw) being patched into one another and thought of leakage, scar tissue, and what would happen to the narrowed area. Would it close completely? I also thought of the mechanical clamps pulling the ribs back after the breastbone had been sawed open, and I thought of my heart being immobilized while the operation was performed—would it start pumping again?"

He decided to go on pills to dilate the blood vessels and pills to keep his heartbeat slow. He took Inderal and Isordil, four times a day. Then he heard of Nathan Pritikin's program of diet and exercise. However, Kimball couldn't afford to take twenty-six days off from work and spend $6000 for a trip to Pritikin's Santa Monica, California, Longevity Center. Instead, he went on the program himself at home.

Kimball eliminated all fat, sugar, caffeine, and nicotine from his diet. He ate mostly complex carbohydrates, including fresh vegetables and cereal grains, and he added an exercise program of walking and jogging. The problem, Kimball found, was that pain would hit his heart just 200 yards from his home. Strength of will and determination to avoid the bypass surgery caused him to push for more distance. He lost some weight, stopped smoking altogether, and managed to make half a mile with half the pain. Finally, at the end of August 1977 he was running and walking four miles at least five times a week. A stress test administered then saw him reach an elevated pulse rate of 178—maximum for his age—and a sustained effort for eight and a half minutes without angina pain. The patient had taken himself off all medication several months before. He no longer needed a bypass.

For the month of May 1978, Kimball gave himself a treat in the Great Smoky Mountains National Park in Tennessee, where he rented a cabin to engage in the "poor man's Pritikin program." He loaded his car with hiking clothes and provisions for the diet against heart disease. Most important were his daily hikes of eight to twelve miles a day in four to six hours over mountain trails. One or more of four physiological processes took place in the arteries to his heart: (1) Their deposits of atherosclerotic plaques blocking the interior vessels were reduced in size, (2) metastatic calcium moved across cellular membranes and out of the smooth muscle cells surrounding each artery, (3) the arteries enlarged as a result of his exercising, and (4) his heart's demand for more blood had lessened, new arteries had formed (collateral circulation), or his blood had become thinner and less viscous (like weight 20 motor oil instead of weight 50) due to the low-fat diet, so that it could more easily pass through narrowed blood vessels.

The patient's next stress test lasted 9.5 minutes without pain and only moderate fatigue in the legs. His pulse went to 90 percent of maximum, and he could have paced longer on the treadmill. Kimball's poor man's Pritikin diet and exercise program had greatly helped to bring his heart muscle back to near normal. While he took an awful chance with his self-administered therapeutic regimen—his heart condition requiring professional supervision—Kimball provided himself with a natural chelation mechanism, as I shall now explain.

Among the needs of the human body are two important factors: the bread of life and the breath of life. Every organ, muscle, and cell depends on a plentiful supply of food and oxygen in the bloodstream. When the fuel line of the body, the vascular system, becomes narrowed or obstructed, food and oxygen are denied to the body's component parts. The vascular system, indeed, is like a city water supply system. After many years of service, the city water system becomes corroded, obstructed by mineral deposits and rust, and deteriorated. If this system could be cleansed or "chelated" by some method, the use and service could be extended considerably.

The human body has such a cleansing mechanism. It is the movement of muscles which keep areas active and supplied with oxygen-carrying blood. Muscle movement massages and carries off the accumulated impurities by stimulating the cells. They give up their wastes to the bloodstream and take in food and oxygen. It's likely that the plaque-forming wastes and metastatic calcium of Rollin M. Kimball would not have narrowed and blocked his coronary arteries so extensively if he had used his muscles more and eaten a proper diet.

What Nathan Pritikin has never acknowledged is that his program of low-fat, high-complex-carbohydrate diet and aerobic exercise actually is partially a natural process of chelation therapy. You don't have to spend $6000 at the Pritikin Longevity Center; you can carry on the poor man's Pritikin program yourself, at home. The exercise program that produces lactic acid, which Pritikin recommends, and the diet foods that are very high in fiber and trace elements that he has people eating do a fine, albeit slow, job of chelating and removing the toxic components that constrict media cells, build plaque along intima walls, trap foreign proteins in between cell membranes, and block blood flow channels. The whole Pritikin program returns to modern man the food and activity of more primitive man, which are natural chelators.

CHELATION SUBSTANCES IN NATURE, COMMERCE, AND MEDICINE

Chelation is a chemical process in nature usually involving a heterocyclic ring compound (a compound containing more than one kind of atom in its ring structure) which binds to or contains at least one metal cation (a positively charged atom) in the ring. Organic compounds in which atoms form more than one coordinate bonding (a shared pair of electrons forms the bond between two atoms) with metals in solution may be chelating agents. These include any weak organic acid such as citric acid, lactic acid, acetic acid, ascorbic acid, and others. Since many food substances have the ability to seize and "sequester" metal atoms such as calcium atoms just like chelating agents, these foods offer a promising foundation for the development of a group of food chelators for the enhancement of health and the prolongation of life. Orthomolecular nutritionists point out that food chelators are the rational and exciting nutritional pharmacology for the future. Although we cannot hope to get as rapid or dramatic effects as we see when the chelator is directly introduced into the bloodstream, we can easily keep up the oral program for a long time and achieve great benefits.

As a mechanism in plant and animal biology, the chelation reaction has been recognized for over thirty years. It is a biochemical process that is fundamental in:

1. food digestion and assimilation
2. the formation of numerous enzymes
3. the functions of enzyme systems
4. the synthesis of many hormones
5. the detoxifying of certain (toxic) chemicals
6. the detoxifying of certain metals such as lead, arsenic, mercury, etc.
7. the movement of vitamins, minerals, hormones and other nutrients across membranes of body tissues as part of transport systems

Chelation is an integral part of human physiology. The process is associated with many familiar substances in nature and in the body. In nature, for example, when magnesium goes to work in plant photosynthesis, it gets chelated by the plant, and the end product is known as *chlorophyll.* In the human body, vitamin B_{12} is a form of chelated cobalt for internal metabolic use. Your body's enzyme cytochrome oxidase contains both iron and copper in a chelated form.

As a way of being utilized in your blood to carry oxygen and make it available to the cells, iron is chelated in the form of hemoglobin. Without this chelation of iron by protein in the body, iron would be very toxic to the tissues.

When you purchase iron and other minerals in a health food store, you will notice the label may claim that they are "chelated" or amino-acid-complexed. The label is referring to a laboratory process that, if properly done, is slightly different from the physiological process occurring in the body. Yet, the laboratory chelation method can accomplish the same desired effect—markedly enhanced absorption and utilization by the body of the inorganic mineral.

Chelated minerals are powerful nutrients that are comprised of a mineral complexed or bonded with a ligand such as an amino acid. Ascorbic acid or aspartic acid can be used, too. A chemical conversion takes place between the ligand and minerals and the newly formed complex helps the body utilize the nutrients in the metabolic processes. For example, in ordinary metabolism, before inorganic iron such as iron sulfate can be absorbed the body must remove the iron from the sulfate (ionize it). Then it must take protein and hydrolize it into amino acids. Finally, through its own chelating processes, the body chelates some of the iron with those amino acids and carries the mineral into the intestinal wall. This natural chelation process generally results in less than 5 percent of originally nonchelated iron being absorbed by your body. (Note: Almost zero is absorbed if you drink tea at the same time, because an insoluble iron tannate complex is formed.) Once the small amount of iron has entered the intestinal wall, your body combines ferric hydroxide-phosphate compound to form *ferritin,* which is the first stage in the absorption of iron.

This resulting iron amino acid chelate carries the iron across the intestinal cells of the bloodstream. The iron is released in the bloodstream and rechelated a third time with a beta-globulin, which is similar to hemoglobin, from the blood. The new chelate, called *transferrin,* moves the iron to the required body stores, such as the liver or the marrow of the bones.[2,3]

Your ingestion of properly chelated minerals is an exceedingly important subject in any prevention program for hardening of the arteries. You can acquire a full range of chelated minerals from almost any health food store or by mail order purchase directly from the laboratories and companies that manufacture and distribute nutritional supplements. However, there are companies that claim their mineral is chelated when it is not. Before taking minerals arbitrarily, have a hair mineral analysis to determine which minerals you need.

In articles written by Harvey Ashmead, Ph.D., and DeWayne Ashmead, Ph.D., Chief Scientists of Albion Laboratories, Clearfield, Utah, they have said, "Certain mineral deficiencies may result in calcification of the heart and arteries . . . . Chelates can be and are produced under laboratory conditions that are almost identical to what the body makes. Because the body does not have to make significant changes in the mineral presented in the properly chelated form, independent laboratory tests have shown, absorption and retention of properly chelated minerals is much higher than non-chelated minerals."[4]

The chelation process may be partly responsible for the activity of many of the drugs used in medicine to treat disease, including aspirin, terramycin, cortisone, adrenaline, penicillin, and a host of others. Without it drugs could not act as effectively and react easily with the body's physiology. Many drugs used in chemotherapy against cancer work through their ability to chelate. Some of the most basic and yet complex chemical reactions found in nature and in man are encompassed by the chelation process.

You can graphically illustrate for yourself how the chelation process works on calcium. Take acetic acid in the form of vinegar and drop some eggshells into it. Watch over several days how the eggshells get thinner and softer and finally disappear, as you add more vinegar to the solution. The weak organic acid of vinegar has chelated calcium out of the eggshells slowly and safely, the way EDTA chelation therapy performs the process for the cells in calcified arterial walls.

As I pointed out in the previous chapter, on EDTA safety, the chemical principle of chelation has been in use for at least a century in many industrial practices. For an example of its use in commerce, look at the softening of water, which is a process of chelation. Certain ion exchange materials, such as zeolite (used to soften water), are chelates. Zeolite is a complex sodium compound. When it comes in contact with hard water, it exchanges sodium for calcium and magnesium, which are the two minerals that make water hard. No lather forms when soap is added to hard water; a scummy precipitate drops out instead. Zeolite chemically forms sodium carbonates and sulfates to provide a softened water. That is exactly how EDTA and other chelating agents work, too.

Chelation also forms the basis for the action of some of the other more commonly used detergents. Just as the chelating physician does for patients, the modern homemaker employs the principle of chelation when she uses detergents to wash clothes and dishes. Detergents form soluble chelates with calcium and magnesium that are readily washed out by water rinsing. That way no scum builds up as a ring around the dishwashing machine, bathtub, or washbasin.

The chelation process is used extensively in modern medicine. For instance, Ringer's lactate, the vehicle I have mentioned previously that is often used for intravenous feedings, contains a natural chelator. It's lactic acid—accounting for an unappreciated therapeutic effect in patients who receive this fluid. Ringer's lactate actually lowers the serum calcium exactly like EDTA. In addition to those I have already referred to briefly, some other pharmaceutical chelates include penicillamine, tetracycline, streptomycin, bacitracin, oxytetracycline, polmyxin, ammonia, cyanide ion, glycine, oxine, dipyridyl, tetraethyllithuram, isoniazid, aminosalicyclic acid, and many other common generic drugs. Each chelating agent has a varying affinity for a variety of cations, depending upon the various spatial arrangement of the bound-up molecule and many additional chemical and physical factors such as the pH or acidity of the solution, etc.

The earliest medical therapeutic use for which the chelation principle was applied was with dimercaprol or British Anti-Lewisite (BAL), discovered during World War II by Professor R. A. Peters and co-workers in Oxford, England. BAL is an antidote against the skin-blistering poison gas lewisite. Chelating the three arsenic atoms in the lewisite molecule renders the gas harmless and easily removable from the skin by water or from the body tissues in the urine. BAL became the first chelating agent used in medical practice. It was the routine treatment for arsenic and other metal poisons during the 1940s. Unluckily, BAL's own irritating effects upon living tissues, and side effects, severely limited its widespread employment. Subsequently, other chelating agents were sought for use internally with fewer undesirable side effects. EDTA became employed for therapeutic purposes because of its lack of side effects, and it's a much better chelator of certain minerals, like calcium and lead, than BAL.

EDTA has been designated by a number of names by physicians, chemists, and other scientists in the papers they have had published. Various references to EDTA have included disodium ethylene-diamide tetraacetate; disodium ethylenediamine tetraacetic acid; edathamil; Endrate; edathamil calcium disodium; Disodium Endrate; Sequestrene; Disodium Versenate; triolene-B; disodium EDTA; and many other labels in foreign countries.

CHELATING YOURSELF AT HOME

Using ethylenediamine tetraacetic acid on yourself at home would be too risky without the medical subspecialty supervision supplied by the participating physicians of the American Academy of Medical Preventics (AAMP). Yet, there are common natural methods of exercise, diet, and nutritional supplementation recommended by orthomolecular nutritionists and holistic physicians that any man or woman can use for accomplishing chelation self-treatment.

While I believe that the exercise and dietary chelation self-treatments set forth in the following pages are beneficial, such self-treatments are not intended to be a replacement for the medical supervision of your personal physician, particularly with respect to the treatment of cardiovascular problems and other medical conditions.

First among the natural chelating methods is aerobic exercise. "Lactic acid is a chelating agent, somewhat comparable with citric acid in selectivity and potency," says gerontologist Johan Bjorksten, Ph.D., in his book on longevity.[5] He explains that lactic acid is formed in muscular actions of any kind. In aerobic exercise, the lactic acid percentage in blood increases to a high level so long as important use of major muscles takes place,[6] as in speed walking, running, rebounding, rope jumping, or any other activity where endurance is stressed, rather than peak performance. Immediately on cessation, the lactic acid content drops back to normal, but while it is elevated natural chelation therapy is taking place.

"Lactic acid is not as effective as EDTA in speed," affirms Dr. Bjorksten, "but given enough time to act, it seems comparable in total removal of chelatable metal. The favorable therapeutic results often obtained by muscular exercise in cases of vascular impairment might well be due to internal chelation even as much as to improvements in blood circulation. This internal chelation would include removal of undesirable deposits of calcium and most of the toxic aluminum, as well as of lead, mercury and cadmium. It may, however, still leave a final residue, greater than one percent of the undesirable metal, to be scavenged by an ultimate chelation using the most potent chelators."

In an earlier article he had published in *Rejuvenation*, Dr. Bjorksten suggested that the beneficial effects of exercise on longevity will be found to be highest in sustainment type activities such as walking, jogging, rowing, skiing, and swimming, where an elevated lactic acid content of blood is maintained regularly and for considerable time. He believes that the chelation process occurring naturally in the body from exercise as well as the intravenous injection of EDTA provide a person with an increased life-span and improved quality of life.

Bjorksten's assertion is that "the lactic acid or lactate content of blood is about doubled during the time of moderate muscular exertion, and declines to the normal level abruptly, in a few minutes upon cessation of the muscular action. Since lactic acid is a fair-to-good chelating agent, it may help remove from the system potentially crosslinking aluminum, cadmium, mercury, lead, arsenic and excess iron, among others, thereby increasing longevity if depletion of the needed chelatable metals manganese, cobalt, zinc, iron and perhaps molybdenum is counteracted by supplemental medication or a couple quarts daily of skim milk."[7]

The duration of your muscular exertion is more important than its intensity, in order to achieve a chelating effect from exercise. Remember that lactic acid buildup is the key to the chelating mechanism in exercise, and lactic acid accumulates mostly after your exercising has been prolonged for a longer period.

THE DR. RINSE FORMULA

Certain food substances are natural chelators. A physical chemist, Jacobus Rinse, Ph.D., of East Dorset, Vermont, has available many case histories which I have seen of people who reversed hardening of the arteries and cured themselves of heart disease. They used a combination of food supplements developed by Dr. Rinse, including his recommended mixture of linoleate, lecithin, wheat germ, brewer's yeast, and several vitamins and minerals. The National Board of Health of Holland has adopted his mixture for preventing heart disease. The proportion of ingredients in the Dr. Rinse formula are:

4 gm lecithin
12 gm coarsely chopped sunflower seed (provides linoleate oil, potassium, and zinc)
5 gm debittered brewer's yeast
2 gm bone meal (dicalcium phosphate)
5 gm raw wheat germ
500 mg vitamin C
100 IU vitamin E
40 mg vitamin B_6
Plus appropriate amounts of magnesium oxide and zinc required by the individual, as indicated by his personal hair mineral analysis

Dr. Rinse's formula should be consumed daily. Make a sufficient supply for a month and keep it refrigerated, using one heaping tablespoonful in yogurt, juice, shakes, cereal, pancakes, or other ways you may invent. Inclusion of the formula's main ingredients is based on research indicating that linoleate-lecithin present in the Lecithin-Cholesterol-Acyl-Transferase (LCAT) enzyme chelates cholesterol deposits on the arterial walls.[8] Thus,

LCAT acts as a form of natural body chelator, and there's logic to eating food supplements that provide more of the enzyme ingredients. Additionally, two medical researchers from the University of Alabama, Butterworth and Krumdieck, came to the conclusion that linoleate-lecithin and vitamin C dissolve cholesterol at normal body temperatures.[9]

You may acquire the Dr. Rinse formula already prepared in one-serving packages from your health food store, or you may contact the Vitaflo Company, William Floyd Parkway, Shirley, NY 11967; telephone is (516) 924-4000.

VITAMIN C

In Chapter 3 I described Dr. Earl Benditt, a cell biologist at the University of Washington, Seattle, and his "monoclonal hypothesis" of why arteries harden. It involves a single smooth muscle cell division similar to tumor formation which takes place in the media. Benditt suggests that as a result of the subdivision, smooth muscle cells crowd into the lumen and reduce the passage of blood. These cells eventually die and lay down cholesterol crystals.[10]

The HDL (high-density lipoprotein) with its increased lecithin content has the capacity to dissolve cholesterol crystals. When the HDL and cholesterol pass through the liver together, HDL is removed and decomposed. Then vitamin C goes to work. The cholesterol is converted into bile acids with the assistance of vitamin C in a type of chelation action.

The human body doesn't make its own vitamin C. You have to supply it from food sources or by taking ascorbic acid supplements. You'll be benefited from experiencing increasing tone and elasticity of blood vessels. The vitamin helps to support the adrenal glands, in an antistress factor, improves iron absorption, converts folactin into folic acid, lowers blood cholesterol levels by the action described, combats many toxins, increases interferon during viral conditions, and achieves many other metabolic functions.

While I believe the dosage of vitamin C recommended by Dr. Rinse is ten times less than the amount you actually need each day for an optimum chelation effect, the vitamin as a nutritional supplement is an excellent chelating agent at any concentration. (But higher doses of ascorbic acid for the chelation effect are better than lower doses.) The usual megadose for the best body response, arrived at gradually over a couple of months, is 4000 mg a day taken with the components of vitamin P—citrin, hesperidin, rutin, flavones, and flavonals—known collectively as mixed bioflavonoids. You can eat large quantities of the following foods, for they are fine sources of vitamin C:

Milligrams (mg) per 100 grams edible portion (100 grams = 3½ oz.)

1300 acerola
369 peppers, red chili
242 guavas
204 peppers, red sweet
186 kale leaves
172 parsley
152 collard leaves
139 turnip greens
128 peppers, green sweet
113 broccoli
102 brussels sprouts
97 mustard greens
79 watercress
78 cauliflower
66 persimmons
61 cabbage, red
59 strawberries
56 papayas
51 spinach
50 orange juice
47 cabbage
46 lemon juice
38 grapefruit juice
36 elderberries
36 liver, calf
36 turnips
35 mangos
33 asparagus
33 cantaloupe
32 Swiss chard
32 green onions
31 liver, beef
31 okra
31 tangerines
30 New Zealand spinach
30 oysters
29 lima beans, young
29 blackeyed peas
29 soybeans
27 green peas
26 radishes
25 raspberries
25 Chinese cabbage
25 yellow summer squash
24 loganberries
23 honeydew melon
23 tomatoes
23 liver, pork

NIACIN

Niacin (vitamin B_3 is a constituent of the coenzymes that take part in transporting hydrogen in protein and carbohydrate metabolism. This is one of the most important vitamin chelators, effecting a reduction of cholesterol deposits and vasodilation in arterial walls. Niacin is a natural blood thinner, as well as being necessary for sugar metabolism, fat synthesis, and tissue respiration. The therapeutic supplement dose of niacin can be from 50 mg up to 3 gm a day, although 50 mg is very useful for most people. One should be cautioned about the severe niacin flush, a feeling of tingling heat in the face and over the body, that accompanies higher dosages of niacin even when divided into smaller equal doses three times a day. If you choose not to supplement with niacin pills, you will ingest this necessary vitamin when you eat some quantities of the following foods.

Milligrams (mg) per 100 grams edible portion (100 grams = 3½ oz.)

44.4 yeast, torula
37.9 yeast, brewer's
29.8 rice bran
21.0 wheat bran
17.2 peanuts, with skins
16.9 liver, lamb
16.4 liver, pork
15.8 peanuts, without skins
13.6 liver, beef
11.4 liver, calf
11.3 turkey, light meat
10.8 liver, chicken
10.7 chicken, light meat
8.4 trout
8.3 halibut
8.2 mackerel
8.1 heart, veal
8.0 chicken, flesh only
8.0 swordfish
8.0 turkey, flesh only
7.7 goose, flesh only
7.5 heart, beef
7.2 salmon
6.4 veal
6.4 kidneys, beef
6.2 wild rice
6.1 chicken giblets
5.7 lamb, lean
5.6 chicken, flesh and skin
5.4 sesame seeds
5.4 sunflower seeds
5.1 beef, lean
5.0 pork, lean
4.7 brown rice
4.5 pine nuts
4.4 buckwheat, whole-grain
4.4 peppers, red chili
4.4 whole wheat grain
4.3 whole wheat flour
4.2 mushrooms
4.2 wheat germ
3.7 barley
3.6 herring
3.5 almonds
3.2 shrimp
3.0 haddock
3.0 split peas

VITAMIN B_6

Pyridoxine (vitamin B_6) exists in the form of pyridoxal phosphate in the arterial cells and functions as a coenzyme (nonprotein enzyme portion) for many different chemical reactions relating to amino acid and protein metabolism. Vitamin B_6 assumes the role of the intermediary between the protein portion of an enzyme known as apoenzymes and enzyme systems. It forms chelates and complexes with various minerals, especially calcium ions, to accomplish the conversions needed in (1) the reaction between amino acids and keto acids, (2) the removal of molecules of carbon dioxide from organic compounds, usually carboxylic acid, (3) synthesis of tryptophan, which is an essential amino acid, and (4) conversion of a number of other amino acids into various substances required by the functioning cells.

The amino acid homocysteine may be responsible for hardening and narrowing the arteries by stimulating the growth of cells along the delicate inner arterial walls. This occurs when there is a deficiency of vitamin B_6, allowing the buildup of homocysteine in the blood. Vitamin B_6 is able to prevent this homocysteine accumulation through the pyridoxine chelation reaction. As a result of proper enzymatic function, vitamin B_{12} and folate coenzymes act to transfer homocysteine into methionine (another amino acid) and then into two molecules of choline. Choline is one of the B-complex vitamins, a lack of which causes fatty liver or cirrhosis. Kidney damage also takes place during a deficiency of choline resulting in high blood pressure. In addition to fat metabolism, this vitamin is involved in body detoxification, neurotransmission, and tissue cell membrane formation.

The interrelationships between the various substances mentioned center on the presence of pyridoxine. Without it in your diet, a series of enzymatic mishaps begin. The recommended megadose intake of pyridoxine in pill form is 250 mg daily, or, if you can, eat a goodly supply of the following foods in which vitamin B_6 is found.

Milligrams (mg) per 100 grams edible portion (100 grams = 3½ oz.)

3.00 yeast, torula
2.50 yeast, brewer's
1.25 sunflower seeds
1.15 wheat germ, toasted
.90 tuna, flesh
.84 liver, beef
.81 soybeans, dry
.75 liver, chicken
.73 walnuts
.70 salmon, flesh
.69 trout, flesh
.67 liver, calf
.66 mackerel, flesh
.65 liver, pork
.63 soybean flour
.60 lentils, dry
.58 lima beans, dry
.58 buckwheat flour
.56 blackeyed peas, dry
.56 navy beans, dry
.55 brown rice
.54 hazelnuts
.54 garbanzos, dry
.53 pinto beans, dry
.51 bananas
.45 pork, lean
.44 albacore, flesh
.43 beef, lean
.43 halibut, flesh
.43 kidneys, beef
.42 avocados
.41 kidneys, veal
.34 whole wheat flour
.33 chestnuts, fresh
.30 egg yolks
.30 kale
.30 rye flour
.28 spinach
.26 turnip greens
.26 peppers, sweet
.25 heart, beef
.25 potatoes
.24 prunes
.24 raisins
.24 sardines
.23 brussels sprouts
.23 elderberries
.23 perch, flesh
.22 cod, flesh
.22 barley
.22 cheese, camembert
.22 sweet potatoes

.21 cauliflower
.20 popcorn, popped
.20 red cabbage
.20 leeks
.20 molasses

VITAMIN E

Vitamin E, taken only as natural—all dextro (D), not levo (L) rotatory—mixed tocopherols, is a chelating nutrient helpful to those with hardening of the arteries. Read the label to make sure you're getting (D) and not (L) vitamin E. It has certain properties important to the body. Vitamin E is an anticlotting agent that improves oxygen utilization, controls the patch of scar forming in damaged heart tissue, and improves capillary permeability. More vital even than these attributes, the vitamin achieves a platelet normalization effect by reducing platelet stickiness, similar to the workings of EDTA. Vitamin E increases prostaglandin X production. Prostaglandins are a hormonelike class of chemicals that control important body functions. Prostaglandin X is manufactured in the artery linings and converts damaged platelets back to normal.

There are only two major pharmaceutical manufacturers of vitamin E in the United States. The Eastman Kodak Co., of Rochester, New York, makes the natural all-dextro (D) form and Hoffman-LaRoche of Nutley, New Jersey, makes a synthetic form, which research indicates is less effective. The biological activity of natural D-alpha-tocopherol is 1.49 IU per mg, whereas synthetic vitamin E is merely 1.1 IU per mg.

Ordinarily the essential fatty acids released into the blood by polyunsaturates are highly vulnerable to peroxidation (burning up), linking their molecules one for one with molecules of oxygen. Vitamin E in the bloodstream, however, preferentially bonds with fatty acids and prevents their oxidation. The vitamin E is destroyed in the course of this activity, which is why polyunsaturates in the diet in any quantity create a need for proportionately more vitamin E. As a result of this antioxidant activity, the vitamin helps to prevent oxygen from being converted into toxic peroxides, leaving the red blood cells more fully supplied with pure oxygen which the bloodstream moves to all tissue cells of the body.

A 1981 report in the *Journal of the American Medical Association* has linked the ingestion of high doses of vitamin E to numerous undesirable side effects, including everything from hypertension to thrombophlebitis. Perhaps I can answer the *JAMA* criticism of this excellent chelating nutrient. I believe that these potential side effects are very rare, and not really from the vitamin E itself. Untoward reactions have been reported from anything taken by mouth, including food, table salt, vitamins, and drugs, and even from placebos. Therefore, I surmise that some reported "side effects" may have not actually derived from vitamin E. They could have come from outside sources such as the capsule or tablet carrying the vitamin E, which may include many excipients, binders, and other substances to hold or carry the tocopherol oil. Also I know that some vitamin E capsules or tablets sold in stores were proven not to contain the stated potency of the vitamin. Occasionally the "E" capsules have been found to contain only a rancid oil. When the oil goes bad, no vitamin E is present in the capsule at all. This circumstance could have harmful side effects that may erroneously be attributed to vitamin E.

You should always discontinue taking any supplement if an untoward reaction develops. Consult a nutritionally oriented physician to help determine what is the cause of your problem. Happily, the AAMP physicians rarely encounter such difficulties in medical practice, probably because of their knowledge of the best supplement sources. They use a broadly based nutritional support program, rather than just using high doses of any one vitamin such as vitamin E.

You can supplement your diet with 100 to 400 to 800 IU daily of natural, dextro-rotatory vitamin E for its chelation effect. If you have some objection to taking a nutritional supplement, try to get part of your vitamin E supply by eating lots of the following foods. Note: Vitamin E is up to ten times more effective when adequate selenium is provided in the diet and/or with supplements.

IU per 100 grams edible portion (100 grams = 3-1/2 oz.)

IU	Food
216.0	wheat germ oil
90.0	sunflower seeds
88.0	sunflower seed oil
72.0	safflower oil
48.0	almonds
45.0	sesame oil
34.0	peanut oil
29.0	corn oil
22.0	wheat germ
18.0	peanuts, raw
18.0	olive oil
14.0	soybean oil
13.0	peanuts, roasted
11.0	peanut butter
3.6	butter
3.2	spinach
3.0	oatmeal
3.0	bran
2.9	asparagus
2.5	salmon
2.5	brown rice
2.3	rye, whole
2.2	rye bread, dark
1.9	pecans
1.9	wheat germ
1.9	rye and wheat crackers
1.4	whole wheat bread
1.0	carrots
.99	peas
.92	walnuts
.88	bananas
.83	eggs
.72	tomatoes
.29	lamb

DIMETHYLGLYCINE

Formerly, and incorrectly, known as vitamin B_{15}, a "metabolic enhancer" known as N.N.-Dimethylglycine (DMG) taken in a dosage of 90 to 270 mg a day provides an internal chelation effect. DMG has the ability to increase available oxygen and thus may provide useful activity in protecting oxidative processes in the body's cells and tissues. This action is associated with the methyl groups in the product's formula, which are oxidized to formaldehyde.[11] The transformation of DMG for use by the body as a metabolic enhancer and chelating agent is carried out by the relevant enzyme systems localized in the mitochondria of the liver. Mitochondria are structures occurring in varying numbers in the cytoplasm of every cell, which is the site of the cell's energy production. Mitochondria provide the source of energy and contain many of the enzymes involved in the cell's metabolic activities.

Various forms of atherosclerosis such as coronary artery disease with manifestations of acute or chronic insufficiency of the coronary circulation or cardiovascular insufficiency, obliterating hardening of the arteries of the lower limbs, and even partial occlusion of the brain's blood vessels may all be benefited from your dosing yourself with DMG supplements each day. It's available in powder form or in tablets, and 180 mg per day is an average dose. You can also ingest some DMG by eating large quantities of the foods in the list that follows. No portion size is given because DMG varies in food according to the soil conditions in which it grows or the food a farm animal eats.

apricot kernels
yeast
liver
rice bran
wheat bran
wheat germ
sunflower seeds
pumpkin seeds
oat grits
corn grits
whole grain cereals

SELENIUM

The mineral selenium is an antioxidant and may also work as a form of chelator by removing free radicals. It helps the body assimilate and utilize vitamin E. Free radicals are harmful atoms or molecules floating around in the body whenever you are exposed to stressors. Selenium is a scavenger for injurious free radicals and helps remove them. Physicians practicing metabolic cancer therapy also have found that mineral is useful in preventing cancer and heart attacks.

Selenium works to retard protein missynthesis. Missynthesized proteins are treated as foreign components by the body. Gerontologist Alex Comfort, M.D., in his 1970 *Gerontologica* paper describes the chelation action of selenium "like protecting a phonograph record by lubricating the needle to reduce scratching with use that would make it unplayable."

The usual dosage of selenium you might take as a supplement is 100 to 200 mcg. If you prefer not to take it in pill form, you can find selenium in the following foods:

Micrograms (mcg) per 100 grams edible portion (100 grams = 3½ oz.)

146 butter
141 smoked herring
123 smelt
111 wheat germ
103 Brazil nuts
89 apple cider vinegar
77 scallops
66 barley
66 whole wheat bread
65 lobster
63 bran
59 shrimp
57 red Swiss chard
56 oats
55 clams
51 king crab
49 oysters
48 milk
43 cod
39 brown rice
34 top round steak
30 lamb
27 turnips
26 molasses
25 garlic
24 barley
19 orange juice
19 gelatin
19 beer
18 beef liver
18 lamb chop
18 egg yolk
12 mushrooms
12 chicken
10 Swiss cheese
5 cottage cheese
5 wine
4 radishes
4 grape juice
3 pecans
2 hazelnuts
2 almonds
2 green beans
2 kidney beans
2 onion
2 carrots
2 cabbage
1 orange

MAGNESIUM

Magnesium nutritional supplementation is a marvelous way to acquire chelation therapy for yourself, since it displaces calcium within the cells. Magnesium alone, if taken by the American population, according to Mildred Seelig, M.D., Ph.D., would cut the heart attack death rate each year from 54.7 percent of everyone who dies to less than half that percentage. The mineral automatically stops arterial spasm in the same way as the calcium channel blocking agents work. It displaces calcium out of the cells in the walls of the arteries. Supplementation with magnesium tablets would require about 1000 mg daily. On the other hand, you may ingest it from the following foods:

Milligrams (mg) per 100 grams edible portion (100 grams = 3½ oz.)

760 kelp
490 wheat bran
336 wheat germ
270 almonds
267 cashews
258 blackstrap molasses
231 brewer's yeast
229 buckwheat
225 Brazil nuts
220 dulse
184 filberts
175 peanuts
162 millet
160 wheat grain
142 pecans
131 English walnuts
115 rye
111 tofu
106 beet greens
90 coconut meat, dry
88 soybeans, cooked
88 spinach
88 brown rice
71 dried figs
65 Swiss chard
62 apricots, dried
58 dates
57 collard leaves
51 shrimp
48 sweet corn
45 avocado
45 cheddar cheese
41 parsley
40 prunes, dried
38 sunflower seeds
37 common beans, cooked
37 barley
36 dandelion greens
36 garlic
25 raisins
35 fresh green peas
34 potato with skin
34 crab
33 banana
31 sweet potato
30 blackberry
25 beets
24 broccoli
24 cauliflower
23 carrot
22 celery
21 beef
20 asparagus
19 chicken
18 green pepper
17 winter squash
16 cantaloupe
16 eggplant
14 tomato
13 cabbage
13 grapes
13 milk
13 pineapple
13 mushroom
12 onion
11 orange
11 iceberg lettuce
9 plum
8 apple

ASPARTATES AND OROTATES

When supplementing with magnesium, it's recommended by holistic medical experts who treat disease with nutrition that you consider the aspartate and orotate forms or perhaps a combination of these, particularly if you are already ill. However, due to expense, many people will have to settle for less sophisticated nutrients such as magnesium alone, amino acid chelates, or even magnesium oxide. These can still offer important protection and may be very useful for the generally healthy person who is primarily using magnesium supplements to help balance and optimize his mineral intake.

Aspartic acid, especially L-aspartate, goes into the inner layer of the outer cell membrane and releases the magnesium, thus helping displace the calcium ion that is attempting to permeate the cell wall. L-aspartate is a true chelating substance, derived from the essential amino acid—aspartic acid. It is often combined in a product containing magnesium L-aspartate and potassium L-aspartate together. One source of this material in the United States is OCI Dietary Products, 4881 Topanga Canyon Blvd., Woodland Hills, CA 91364; (213) 884-2660. Using the magnesium supplements such as the aspartates, you may experience an increase in your heart pump blood volume and an increase in total energy because magnesium is an activator for the metabolism in cells so it helps them perform additional work. This is an important food supplement for athletes wanting maximum performance during their competitions, as well as for patients with heart disease, including complications from atherosclerosis or vascular spasm.

Orotic acid, first isolated from milk whey has been shown to play a useful role in active mineral transport. The orotates are not just an efficient form of magnesium supplementation. Their chelating effects have shown benefits against hardening of the arteries and are partially related to the activity of the orotic acid barrier molecule. It has a diversity of roles in metabolism, which include:

1. Stimulates oxygen-dependent pentose pathway tissue such as bone, cartilage, blood vessels, heart and liver mesenchyme.
2. Promotes a body-building effect that increases protein synthesis necessary in the restoration of aging tissue.
3. Supplies building blocks for nucleic acids (RNA and DNA) synthesis, which control every aspect of body metabolism.
4. Normalizes bile flow.
5. Activates white blood cells and their digestive capacity.

Magnesium orotate and/or potassium orotate can be taken alone or together for their beneficial chelation-like effects in place of or in conjunction with the aspartates. These nutrients work best for long-term therapy as chelating-type of agents in the self-treatment or prevention of hardening of the arteries.

Using high doses of the orotates of not less than 1.5 gm magnesium orotate, 150 to 300 mg potassium orotate, and 120 to 400 mg bromelain, Hans Nieper, M.D., Chief of the Department of Medicine, The Silbersee Clinic, Hanover, West Germany, was able to reduce reinfarcts (additional degeneration and scarring) in heart attack patients from 34 percent to less than 2 percent over an observation period of four years.[12]

Magnesium orotate is transported into the cells of blood vessel walls, where magnesium is released to act as a calcium displacer. No other magnesium product has been known to work quite as effectively against cell proteins binding calcium. Magnesium orotate markedly lowers blood fat levels; it functions in carbohydrate metabolism and has a normalizing effect on blood clotting; it activates liver enzymes in glucose phosphate metabolism, which affects the cells of the artery walls. Magnesium released locally affects the smooth muscle coat of the blood vessels and reduces elevated blood pressure.

Reports of liver toxicity due to high doses of magnesium orotate have appeared in the literature. Liver problems in their patients using orotates have not been reported by the AAMP physicians, who always employ a broad-spectrum nutritional program. They never rely on just vitamin E or just vitamin C or just selenium or just magnesium. This broad-spectrum approach to nutritional supplementation means spending more money and the inconvenience of swallowing more pills. But your action appears to be well justified, as the incidence of adverse side effects from the basic nutritional program outlined in the Protocol and followed by AAMP physicians seems to be virtually nil.

Nutritionists know that extra choline and inositol protects against potential niacin toxicity. I believe many more of such protective actions will be uncovered in the future, further justifying the broad spectrum supplementation approach. There is nothing wrong with taking supplemental nutrition as long as you check what you're doing with a knowledgeable doctor who uses orthomolecular nutrition as a therapeutic and preventive tool for his or her patients.

Please note that knowledge of the aspartates and orotates will probably be rather limited even among some physicians who use nutrition as a therapeutic tool. Don't be surprised if your average local doctor doesn't know what you're talking about if you ask him about any of these self-treatment chelates to enhance the health of your cardiovascular system.

MANGANESE

The typical American diet, with about 60 percent of its calories from refined sugar, refined flour, and fat, most of which is saturated, is designed not only to provide as little manganese as feasible, but also to cause depletion of body stores of chromium by not replacing urinary losses. But manganese is a necessary natural chelating agent when taken in the food supply. Authorities at Columbia University College of Physicians and Surgeons state that manganese will do everything that the calcium channel blocking agents do but more naturally and effectively. It is a major factor in inhibiting calcium from entering the cells lining the arteries.

An excess accumulation of calcium within the cells is now agreed by most cardiovascular experts to bring on degenerative disease. Six million spontaneous bone fractures occur from osteoporosis, for instance, from calcium leaching from the bones and teeth into the soft tissues. Using manganese, however, by taking it in our food or as a supplement, is the same as getting chelated constantly. Providing the cell with manganese and keeping it inside by eating foods with manganese helps to hold off calcium from invading the cell. There will be no intracellular infiltration by calcium transporting through the cell membrane as long as manganese is present as a substitute. It works in a similar way to magnesium's characteristic of displacing calcium ions.

The practice of liming the soil in America causes manganese not to be bio-available. Lime keeps manganese out of crops because of a change in ionic balance. It gets replaced in soil by phosphorous and calcium by chemical manipulation of the farmers. Additionally, Americans compound their problem of manganese deficiency by taking a tremendous quantity of phosphorous into their bodies in the form of carbonated soda waters (phosphates) and a disproportionate amount of meat in the diet. Colas and condensed milk contain too much phosphates. Beef is currently forty-four times higher in phosphorus than 100 years ago. Manganese, a natural form of chelation therapy when ingested, is being eliminated from the American diet.

The more primitive people in the Philippines and in Vilcabamba, Ecuador have much higher manganese content in their hair mineral analyses than do any persons in the industrialized West. The primitives have a much more normal calcium-phosphorus ratio than western societies; therefore, they don't have hyperparathyroidism to send calcium into the cells and prevent manganese protection. These factors hold off the incidence of hardening of the arteries among these people.

You should attempt to acquire more manganese in your diet. Taking a manganese food supplement of 5 to 25 mg manganese per day will not markedly elevate your manganese hair analysis, but it will increase it slightly. A better practice would be to eat whole foods containing manganese in higher quantities.

Milligrams per 100 grams edible portion (100 grams = 3½ ounces) of foods containing manganese are:

3.5	pecans	0.13	Swiss cheese
2.8	Brazil nuts	0.13	corn
2.5	almonds	0.11	cabbage
1.8	barley	0.10	peach
1.3	rye	0.09	butter
1.3	buckwheat	0.06	tangerine
1.3	split peas, dry	0.06	peas
1.1	whole wheat	0.05	eggs
0.8	walnuts	0.04	coconut
0.8	fresh spinach	0.03	apple
0.7	peanuts	0.03	orange
0.6	oats	0.03	pear
0.5	raisins	0.03	lamb chops
0.5	turnip greens	0.03	pork chops
0.5	rhubarb	0.03	cantaloupe
0.4	beet greens	0.03	tomato
0.3	oatmeal	0.02	whole milk
0.2	cornmeal	0.02	chicken breasts
0.2	millet	0.02	green beans
0.19	Gorgonzola	0.02	apricot
0.16	carrots	0.01	beef liver
0.15	broccoli	0.01	scallops
0.14	brown rice	0.01	halibut
0.14	whole wheat bread	0.01	cucumber

Cloves, ginger, thyme, bay leaves, and tea are also high in manganese.

BROMELAIN

As a result of some of its actions, bromelain, an enzyme derived from pineapple, provides a chelationlike effect. It acts like a "pipe cleaner" for the blood vessels throughout the entire circulatory tree. Bromelain helps to prevent the narrowing of coronary arteries that contributes to angina and heart attacks. The enzyme reduces the incidence of inflammation that comes with phlebitis. It safely assists in thinning the blood, delaying red blood cells in their clumping action, and lowers the risk of blood clots. Bromelain also selectively increases tissue cell permeability so that other nutrients taken orally can get into the tissues faster. Bromelain is found in most fresh tropical fruits such as pineapple, papaya, and mango. You can also supplement with tablets acquired from the health food store.

PURIFIED DEORDORIZED GARLIC Among the best natural chelators available is an herb—garlic—that has been widely used in worldwide medicine for years. Garlic is specially processed now by the Japanese to remove its odor. It is grown in special gardens far from automobiles to assure that it is free of dangerous contamination from heavy metal pollution such as lead. Nothing in its processing inhibits garlic's effectiveness.

Kyolic, a purified form of this special garlic, is produced by Wakunaga of America Co., Ltd., 23510 Telo Avenue, Suite 5, Torrance, California 90505; telephone (800) 421-2998. It comes from cultivation in rich soil that is also completely free of chemical fertilizers and insecticides.

Kyolic is aged for twenty months, a natural way to metabolize garlic's odor-producing agent. Aging concentrates its chelating ability to 200 times the effect mildly furnished in the fresh garlic clove. The herb's potentially toxic factor, allicin, is reduced from Kyolic, leaving it as purified garlic with specific therapeutic properties. Kyolic garlic may safely be taken by anyone in amazingly large doses, producing the following actions:

1. Purified garlic is a type of antibiotic, which clears local infections such as boils, and even benefits sore throats, upset stomachs, food poisoning, flu, etc.
2. Purified garlic tends to increase your energy by elevating the level of adenosine triphosphate (ATP). This fights fatigue and enhances athletic performance.
3. Purified garlic catalyzes the body's utilization of vitamin C. Blood levels of ascorbic acid are elevated for longer periods.
4. Purified garlic guards against the effects of radiation.
5. Purified garlic chelates heavy metals such as lead and mercury, allowing them to precipitate and be removed from the body through excretion. (It has been shown to increase the fecal excretion of mercury as much as 400 percent and to completely protect blood cells against high levels of lead.)
6. Purified garlic prolongs the beneficial effects of vitamin B_1.
7. Purified garlic contains large amounts of selenium, germanium, and sulfer, which are clinically useful trace elements: for example, selenium is exceedingly useful in protecting against hardening of the arteries and cancer.

8. Purified garlic spares the antioxidant properties of vitamin E.
9. Purified garlic stimulates the liver mitochondria and thus provides an antidiabetic action.
10. Purified garlic acts to maintain the integrity of red blood cells and is thus an antianemic factor.
11. Purified garlic accelerates protein synthesis.
12. Purified garlic increases spermatogenesis so that men experience greater potency. It has also been found useful in the elimination of sexual apathy and stimulation of the libido.
13. Research indicates garlic is useful to help reverse and/or prevent the development of arteriosclerosis.
14. Numerous other useful actions of garlic are being discovered almost daily.

Try the following purified garlic drink:

Juice of ½ lemon
¾ cup lukewarm water
1–2 teaspoons honey
1 teaspoon Kyolic
1–2 tablespoons sweet wine (optional)

Mix well and drink.

The purified product comes in several forms, including liquid garlic extract, tablets, and capsules. The recommended dosage for cardiovascular health is two 270 mg tablets or two capsules of garlic extract powder at each meal. However, it is very safe, and large doses of as many as twenty pills at once can safely be taken when the need exists.

ONIONS

Something in onions provides a chelating effect. A British study performed by Dr. A. S. Truswell, Professor of Medicine at the Queen Elizabeth College of London University, found that eating onions with a high-fat meal reduces a person's chances of having a heart attack or stroke.

Dr. Truswell explained: "Our studies involved nine subjects (in 1978). We served an ordinary breakfast, one that was high in fat and a third one that was high in fat but with about two and a half ounces or three small onions added to it. What we found was that after the fat meal, the subjects' blood platelets stuck together faster than the platelets of the subjects who had eaten the ordinary meal. But when onion was included in the high-fat meal, this effect was neutralized.

"Platelets are a component in the blood concerned with stopping bleeding," explained Dr. Truswell. "When their function goes wrong, they form clots which are involved in thromboses in arteries to the brain and heart—resulting in heart attacks and strokes. If a person is just on the verge of having some sort of blockage of an artery and something increases the stickiness of these platelets such as a high-fat meal, it just might predispose him to a heart attack or stroke."

As with garlic, onions tend to decrease the clotting power of blood. Work performed by Dr. Truswell confirms this. Adding onions to your diet, especially when you consume a fatty meal, will have a chelation effect take hold. The clotting power of the blood is lessened, and you decrease your chances of furthering the hardening of your arteries, at least with eating that particular high-fat meal.

HIGH-FIBER FOODS

High levels of blood fats are associated with increased risk of hardening of the arteries, and high-fiber foods such as those included in the varieties of complex carbohydrates can reduce the amount of fatty substances—cholesterol and triglycerides—in the blood. The soluble fibers such as pectins and gums are most effective as chelators of blood fat. Serum cholesterol reductions of up to 22 percent are achieved simply by adding foods rich in high methoxy-pectins and other soluble fibers, which are present in whole oats, carrots, whole fruits, especially apples, and other roughage to the diet.

The following high-fiber foods should be eaten in quantity, since they are excellent chelating or sequestering agents:

Cellulose, including coarse bran, unpeeled apples and pears, Brazil nuts, fresh peas and carrots.

Hemicellulose, including bran cereals, whole grain breads, beets, eggplant, and radishes.

Lignin, including pears, toasted whole-grain breads, boiled potatoes.

Pectin, including bananas, oranges, apples, potatoes, cabbage, carrots, and grapes.

Gums, including oatmeal, sesame seeds, beans, bulk laxatives.

The main purpose of the noninvasive nutritional supplements and foods described in this chapter is to dramatically improve circulation and control or reverse diseases of the cardiovascular system in a way similar to EDTA chelation therapy. The effectiveness of one or more of these nutrients and of your own lactic acid produced by exercise cannot be evaluated solely on the basis of the chemical affinity between the nutrient and its target calcium ion. You must also consider how readily the nutrient can reach its target. Sometimes the nutrient must penetrate a layer of fatty molecules to make contact with the calcium; in such cases it is essential that the food molecule is fat-soluble. Vitamin E is soluble in fat, but some of the other nutrients are not. What makes them more effective chelators over some other chelating-type ingredients is that they have no bulky molecule groups that will prevent them from penetrating a cellular protein that holds the calcium ion. Moreoever, these nutrients aren't readily inactivated or destroyed by enzymes acting on them in the body. Since the individual nutrients have food value for the body in and of themselves, they do no injury but only enhance your health.

The information in this chapter may seem overwhelming if you are uninitiated into holistic health practices and nutritional therapy. If you have any questions, you should consult a health professional who uses nutrition as part of his practice. The natural chelating foods and supplements supplied here in lists and dosages can go a long way in providing longevity and improving the quality of your life. But you have to use them in a sensible daily program.

CHAPTER EIGHT

Disease Prevention with Chelation Therapy

The investigation of effective chelating foods and drugs is turning u valuable clues to the roles and locations of trace elements in the huma body. The study of chelates is also providing medical scientists with infor mation about the nature of various diseases and basic principles for th design of new treatments. Ideally, we will eventually have ways available t. induce our bodies to produce their own internal curative chelating agents. This would be a major advance in eliminating heavy metal toxicity as a threat to health throughout the industrialized West.

LEAD

It's known, for instance, that a person living in one of the more advanced Western countries in 1982 has 500 to 1000 times the quantity of lead in his body as did his counterpart when Jesus walked on earth. Acute metal poisoning annually afflicts over 2500 adults in the United States. With our present technology spewing enormous quantities of metals into the environment, all of us carry loads of toxic metal in our bodies.

We are exposed to air filled with the combustion of leaded gasoline and to deposits of this heavy metal in our food and water. Clair Patterson, Ph.D., of the Massachusetts Institute of Technology, has estimated that the average adult American takes in more than 400 mcg of lead a day from industrial or household sources and carries a constant load of 200 mg of lead in his body. This toxic load keeps the individual at a subclinical stage of disease. For example, 40 percent of the population of Roxbury, Massachusetts, a suburb of Boston, is affected by lead poisoning. The poisoning comes from lead leaching out of household water pipes. The plumbing is over 100 years old in some cases, and Roxbury's highly acidic water dissolves the metal into the drinking water. The residents could reduce their lead intake almost 50 percent by merely opening their taps for two minutes in the morning before drinking the water after it has stagnated all night in their pipelines.

Acute attacks of lead poisoning occur in slum dwellings where peeling and flaky lead paint is ingested by children. A slum area in Cleveland is known as the "lead belt"; 85 percent of the acute lead poisonings in that city occur in this area. The same is true for the Bedford-Stuyvesant section of Brooklyn, New York. Urine and hair examinations conducted among children living in old, substandard housing in the Cleveland lead belt showed that ten times as many of them suffered from dangerous levels of lead poisoning as did children who lived in a new housing project constructed in the same slum area.

If all inner-city children around the country were tested by hair, blood, and urine analysis, probably one-third of them would be found with acute overexposure to lead. They require EDTA chelation therapy and, in fact, over 100,000 American children have already received the EDTA treatment for lead poisoning at poison control centers. Low-level or chronic lead poisoning is now recognized to exist in all children, to a greater or lesser degree, and can be shown to be producing adverse effects in their behavior and learning ability. The above information is supplied by Herbert Needleman, M.D., Chief of Psychiatry at Harvard University Medical School and the Chairperson for the Lead Committee of the Centers for Disease Control of the United States Government. In a presentation Dr. Needleman made in 1981 to the American Association for the Advancement of Science, he said, "There is no safe level of lead for children in the United States today."

The International Atomic Energy Agency in Vienna, Austria, has fifty scientists from twenty countries participating in a hair analysis research project, checking for and monitoring heavy metal poisoning around the world. The Agency reports that while everyone is worried about the potential nuclear holocaust, the real biological time bomb is poisoning from such toxic minerals as lead. Heavy metal toxicity is predicted by many environmental experts to be the source of the eventual destruction of mankind on earth.

This International Atomic Energy Agency announcement was followed by a shocking study performed over a ten-year span at the University of Miami. Sperm counts recorded on healthy young male students indicate that nearly one-third of single men between ages sixteen and twenty-four will in the future be unable to impregnate their wives. Their sperm counts have dropped over the last eight years from 150 million spermatozoa per cubic centimeter to less than 60 million per cc. Some fertility experts attribute this high rate of male sterility to potential overexposure to toxins in our environment. Environmental toxins definitely include low-level lead poisoning and other forms of heavy metal toxicity.

How do we come in contact with so much lead toxicity? You've already read here that significant sources of increasing your lead exposure come from lead water pipes, canned foods, chipped paint, leaded gas, and industrial pollution. Although these sources are often cited as the main ones for acute lead poisoning, they are not the only sources for chronic lead toxicity.

The increasing prevalence of lead as an environmental contaminant is causing most of us, especially children, to suffer subclinical adverse health effects, said the Natural Resources Defense Council in a report it issued in May 1978. Some serious symptoms and signs of chronic lead toxicity are finally becoming evident. Depression of your immune system is one example. From entrance into the body by inhalation and ingestion, children may absorb up to 50 percent of their lead exposure. Adults will absorb 5 to 25 percent of ingested or inhaled lead, depending on the lead particle size and one's tolerance to the poison. Here are a number of the other signs and symptoms overexposure to lead will produce:

headache
depression
insomnia and/or drowsiness
fatigue
nervousness
irritability
dizziness
confusion and disorientation
anxiety
muscle weakness and wasting
saturnine gout (*saturnine* means symptomatic of lead poisoning)
aching muscles and bones
abdominal pain

loss of appetite
loss of weight
constipation
hypertension
kidney function defects
reproductive defects
  decreased fertility in men
  spontaneous abortion in women
adrenal gland function impaired
iron-deficiency anemia
blue-black lead lines near base of teeth

Symptoms More Common in Children:
hyperactivity
temper tantrums
withdrawal
frequent crying for no apparent reason
fearfulness
refusal to play
other emotional or behavioral problems
drowsiness
learning disabilities
speech disturbances
perceptual motor dysfunctions
mental retardation
seizures or convulsions
ataxia
encephalopathy

Some of the many sources of toxic lead in your immediate surroundings are the following:

atmospheric lead
  motor vehicle exhausts
  lead smelters
  coal burning
  refining lead scrap
  burning materials containing lead
dust and dirt
leaded house paint
drinking water
lead plumbing
vegetation growing by roadside

vegetables grown in lead-contaminated soils
canned fruit and fruit juice
canned evaporated milk
milk from animals grazing on lead-contaminated pastures
bone meal
organ meats, especially liver
lead-arsenate pesticides
wine (leaded caps)
rainwater
snow
improperly glazed pottery
painted glassware
pencils (paint)
toothpaste
newsprint
colored printed materials
spitballs
eating utensils
curtain weights
putty
car batteries
cigarette ash
tobacco
lead shot
mascara
painted children's toys
PVC containers
canned pet food
hair dyes (progressive darkeners)

If the human body were supplied with an effective way of ridding itself of heavy metal toxicity, humankind would hardly be in such fearful danger of destroying itself by means of degenerative disease resulting from industrial pollution. Treatment with one of the various nutritional supplements, foods, or solutions that chelate is almost the only way human beings will be preserved against heavy metal poisonings that often are the cause of degenerative diseases. We will be hearing more frequently of these poisonings as people are exposed to greater environmental pollution. Heavy metal toxicity is part of the reason we in the West experience premature old age from degenerative diseases.

Degenerative disease is not at all a mysterious, genetically regulated, preordained destiny that we must live through and must accept. It is rather a controllable ameliorating process determined by your own lifestyle and the extent of your surrounding ecosystem destruction. You can dictate the outcome of your vitality that you take into the later years of life. The first thing to do is get rid of those vested interests that foist heavy metal intoxication upon you.

Commercial interests are not necessarily confined to the mining industry that digs up deposits of metal, nor are they limited to large manufacturers utilizing heavy metal raw materials for consumer products such as mercury thermometers, aluminum cookware, or lead batteries. Instead, metal pollution may be stimulated by the little guy who doesn't understand or doesn't care that he must be the eventual victim of his personal ignorance on the subject of heavy metal poisoning.

It's the little guy who gets lead poisoning from working in a cannery, gasoline refinery, on the police firing range, as a house painter, or at a tollbooth on the highway. He's the one to lose his job when some interruption in the status quo of heavy metal commerce occurs. If the government tightens regulations and lowers the level of lead to which it allows the average American worker to be exposed in factories using lead, such as those making lead batteries, a rebellion could occur in the battery (or other involved) industry.

The manufacturer will probably decide, "I can't afford to make batteries in the United States anymore. If you bureaucrats force me to clean the air in my plant that much, I will import my batteries from Hong Kong or Afghanistan. No company is going to spend $10 million to make this plant safe for workers."

Then the lead workers realize, "We're going to be out of our jobs. Damn you, government, let us work. We don't care if we have lead poisoning. It's not certain we'll die sooner from lead, but we sure as hell could starve to death." Laborers put pressure on the politicians and win the right to stay on their jobs. The plant—toxic to people—continues to manufacture lead batteries. Conditions remain nearly the same here.

Government officials in this simulated situation have compromised. Their ruling has come down saying that lead levels won't be lowered, but no women of childbearing age should be allowed to work in the battery factory.

The women go out and get hysterectomies. "We're not going to lose our jobs just because we're women," they say. Later, several of these women sue the battery company, declaring that the only way they could keep their jobs was to become sterilized. The ethics, interrelationships, and relevancies of the issues involved with making life safer and healthier take on tremendous significance. They are very weighty issues—as heavy as lead.

Many experts believe the only answer, if mankind is to survive, is to stop all lead mining now, but the gigantic dislocations in the involved industries pose such a staggering economic problem initially that no politician will accept this solution—no matter how convincing the scientific case is—without overwhelming public pressure. Meanwhile, those informed citizens who understand the threat will have to take the necessary nutritional and/or chelation steps I have been describing in prior chapters to protect themselves and their families.

Manufacturing batteries is not the only occupational exposure to lead toxicity, of course, for the list of industrial jobs connected with lead is among the longest in industry. If your income is derived from one of the jobs listed here, you may be in danger of suffering from lead poisoning.

acid finishers
actors
battery makers
blacksmiths
bookbinders
bottle cap makers
brass founders
brass polishers
braziers
brick burners
brick makers
bronzers
brushmakers
cable makers
cable splicers
canners
cartridge makers
chemical equipment makers
chlorinated paraffin makers
chippers
cigar makers
crop dusters
cutlery makers
decorators (pottery)
demolition workers
dental technicians
diamond polishers
dye makers
dyers
electronic device makers
electroplaters
electrotypers
embroidery workers
emery wheel makers
enamel burners
enamelers
enamel makers
explosives makers
farmers
file cutters
firemen
flower makers (artificial)
foundry workers
galvanizers
garage mechanics
glass makers
glass polishers
glost kiln workers
gold refiners
gun barrel browners
incandescent-lamp makers
ink makers
insecticide makers
insecticide users
japan makers
japaners

jewelers
junk metal refiners
labelers (paint can)
lacquer makers
lead burners
lead counterweight makers
lead flooring makers
lead foil makers
lead mill workers
lead miners
lead pipe makers
lead salt makers
lead shield makers
lead smelters
lead stearate makers
lead workers
linoleum makers
linotypers
linseed oil boilers
lithotransfer workers
match makers
metal burners
metal cutters
metal grinders
metal polishers
metal refiners
metal refinishers
metallizers
mirror silverers
musical instrument makers
nitric acid workers
nitroglycerin makers
painters
paint makers
paint pigment makers
paper hangers
patent leather makers
pearl makers (imitation)
pharmaceutical makers
photography workers
pipe fitters
plastic workers
plumbers
printers
policemen
pottery glaze mixers
pottery glaze dippers
pottery workers
putty makers
pyroxylin-plastic workers
riveters
roofers
rubber buffers
rubber makers
rubber reclaimers
scrap metal workers
semiconductor workers
service station attendants
sheet metal workers
shellac makers
ship dismantlers
shoe stainers
shot makers
silk weighters
slushers (porcelain enameling)
solderers
solder makers
steel engravers
stereotypers
tannery workers
television picture tube makers
temperers
textile makers
tile makers
tinners
type founders
typesetters
vanadium compound makers
varnish makers
vehicle tunnel attendants
wallpaper printers
welders
wood stainers
zinc mill workers
zinc smelter chargers

Common nonoccupational lead exposures include:

ceramics, pottery, and related hobbies
stained-glass work
electronics and related hobbies involving extensive soldering
firing ranges
hunting (especially those who cast their own bullets)
eating or drinking from improperly fired lead-glazed ceramic tableware
eating lead-bearing paint
burning battery casings
consuming illicitly distilled whiskey
extensive auto driving (especially in cities)
extensive work with motor fuels
painting with lead-containing paints
home plumbing repairs (lead pipe systems)
exterminating

Many people finally will come down with the symptoms of heavy metal poisoning, since lead levels are still building up constantly in our environment. Chelation therapy remains the single answer for saving the victims from disease, disability, and death. If we truly want disease prevention in our society but industrialization and economics pushes us into despoiling our environment, exercise, specific foods, nutritional supplements, and chelation therapy are the only ways to go.

Heavy metals are among the most dangerous pollutants to which any person exposes himself. Unlike other pollutants, they don't necessarily smell bad, taste awful, look horrible, sound irritating, or feel uncomfortable. In fact, their insidious penetration into our lives in what we eat, drink, breathe, wear, and drive is the great danger. They poison us without warning. And we don't know of the poisoning until the symptoms of heavy metal intoxication show long after the toxins have lingered in our body tissues.

The possible toxic minerals when taken in levels higher than normal ranges include beryllium, antimony, arsenic, mercury, cadmium, lead, aluminum, nickel, copper, tin, fluorine, bromine, vanadium, strontium, iodine, silver, and gold. But keep in mind that any mineral in high enough concentration is toxic. They are listed in no particular order. Some are more toxic than others, and invariably their toxicity varies with the concentration of them in one's environment. In the smallest of traces, these minerals may be tolerated by our bodies and may sometimes be beneficial. Most of the time, however, they are guilty as the underlying killers of 65 percent of everyone who dies on the North American Continent and in Europe.

I have discussed lead toxicity in some detail. This section will continue with a briefer description of three of the other major toxic metals which are thought to promote the process of hardening of the arteries and other degenerative diseases. They include mercury, aluminum, and cadmium.

MERCURY

Mercury is a potent protoplasmic poison, meaning that it is deadly to all living things. Fish have an unusual propensity to accumulate it in their bodies. You eat the fish and take their mercury into your own body. And fish are just one example of how the substance gets into our food chain.

A vinyl chloride and acetaldehyde plant making plastics discharged methylmercuric sulfide and methylmercuric chloride into the Minimata River on Kyushu Island in Japan in 1946 and 1948. The fish and shellfish became polluted with alkyl mercury. They were eaten by the local people, who came down with central nervous system disease, including permanent degeneration of the brain. The government and the medical profession knew of nothing to do for these victims. A person with symptoms of Minimata disease merely received some government money and waited to die. Today, certain new chelators developed in Europe and Japan—but not allowed by the FDA to be used in the United States—hold great promise for this kind of problem.

Mercury poisoning in one form or another comes from ingested organic (methyl) or even inorganic (less toxic) mercury that is absorbed through the gastrointestinal tract. It also derives from inhaled mercury vapor (such as when the dentist is working on your fillings, which he has to inhale all year long!) that is retained by the pulmonary system. Skin absorption of the metal may also occur. Here are the signs and symptoms for which you should be watchful and bring to the attention of a knowledgeable bioinorganic toxicologist who will use chelation therapy to pull the poison out of your body.

Elemental mercury vapor exposure:
insomnia
shyness
nervousness
dizziness
loss of memory
lack of self-control
irritability

anxiety
loss of self-confidence
drowsiness
depression
loss of weight
loss of appetite
tremors

Severe cases:
hallucinations
manic depression

Later findings:
numbness and tingling of the lips and feet
muscle weakness progressing to paralysis
loss of vision
hearing difficulty
speech disorders
loss of memory
incoordination
emotional instability
dermatitis
renal damage
general brain dysfunctions

Late symptoms:
coma
death

Sources of mercury intoxication may be present on your person right now, or they could be in your home or workplace. To evaluate the danger you face in using a product, read the label to see if mercury is one of its ingredients. Here is a listing of the most common items from which mercury poisoning comes:

mercury-silver amalgam (dental fillings)
broken thermometers and barometers
consumption of grain seeds treated with methyl mercury fungicide
fish and marine mammals
mercuric chloride (used in histology labs)
calomel (body powders and talcs)

mercury-containing cosmetics
latex and solvent-thinned paints
organic mercurials (diuretics)
air polluted by industrial mercury vapor
mercury-polluted industrial water
clothing worn by mercury workers
hemorrhoid suppositories using mercurials (such as the most advertised brand name: "Preparation H.")
mercurochrome and thimerosal (Merthiolate)
fabric softeners
floor waxes and polishes
air conditioner filters
wood preservatives
cinnabar (used in jewelry)
batteries with mercury cells
fungicides for use on lawns, trees, shrubs, etc.
tanning leather
felt
adhesives
laxatives (containing calomel)
skin-lightening creams (commonly used by women as a cosmetic to rid themselves of unwanted dark spots or "aging" spots on the skin)
psoriatic ointments
photoengraving
tattooing
lab and industrial equipment using metallic mercury
sewage sludge used as fertilizer
sewage disposal (may release thousands of tons of mercury annually worldwide)

It's not unusual to find mercury intoxication among those who work in dentistry. The suicide rate among dentists is one of the highest of any of the learned professions (practically even with psychiatrists), and mercury poisoning is suspect as the cause. Workman's compensation insurance premiums have been rising for dental offices, because of the high incidence of mercury intoxication in the paraprofessionals assisting the doctor. Take a long hair sample from a young woman after she has worked in a dental office for six months, for instance, and test it for mercury. The end of oldest growth of hair may show no mercury content, but the end of newest growth in the same strand will indicate that mercury is permeating her body. This means that her exposure has occurred during the preceding six months.

The following list shows which jobs expose you to inordinate dosages of mercury, resulting in possible toxicity:

amalgam makers
bactericide makers
barometer makers
battery makers, mercury
boiler makers
bronzers
calibration instrument makers
cap loaders, percussion
carbon brush makers
caustic soda makers
ceramic workers
chlorine makers
dental amalgam makers
dentists
direct current meter workers
disinfectant makers
disinfectors
drug makers
dye makers
electric apparatus makers
electroplaters
embalmers
explosives makers
farmers
fingerprint detectors
fireworks makers
fish cannery workers
fungicide makers
fur preservers
fur processors
gold extractors
histology technicians
ink makers
investment casting workers
jewelers
laboratory workers, chemical
lampmakers (fluorescent)
manometer makers
mercury workers
miners, mercury
mirror makers
neon light makers
paint makers
paper makers
percussion cap makers
pesticide workers
photographers
pressure gauge makers
refiners, mercury
seed handlers
silver extractors
switch makers, mercury
tannery workers
taxidermists
textile printers
thermometer makers
vinyl chloride manufacturing
wood preservative workers

ALUMINUM

In Guam, 12 percent of the population is suffering from symptoms resembling amytrophic lateral sclerosis, which I've referred to previously as Lou Gehrig's disease. They also show symptoms of memory loss and dementia that resemble Alzheimer's disease. There appears to be an endemic aluminum toxicity problem on the island—possibly excessive aluminum was added to the drinking water. The experts do not yet fully agree on exactly what is occurring. It is known that the potential for developing aluminum poisoning has been markedly increased because the residents are now eating the low-calcium, low-magnesium, high-phosphorus diet that is so typical in our Western culture. Dr. Bjorksten has pointed out that future gerontologists wishing to forestall or prevent the aging process will have to monitor aluminum levels in their patients.

Aluminum is the most prevalent mineral on the earth's surface, but we have not been able to find any use for it in the body. Over thirty published articles now link excess levels of aluminum in the brain to premature senility and loss of memory. EDTA chelation therapy takes up or chelates with the trivalent aluminum ion poorly, but some of it could be removed by the chelation method. This could help to prevent us from becoming prematurely aged and ill. Currently the citizens of Guam are dying at a younger age than their parents did and the average life-span is declining.

Sources of potential aluminum toxicity are scattered throughout the Western way of life. Aluminum sources could present a significant health risk—which is naturally disputed by the aluminum industry. The following signs and symptoms may indicate damage to your body tissues from excessive aluminum:

- aluminum pneumoconiosis (inhalation of aluminum dust)
  - pneumothorax
  - pulmonary fibrosis with emphysema
  - dyspnea
  - right-sided cardiac hypertrophy
  - Shaver's disease: cough, substernal pain, weakness, fatigue, bilateral lacelike shadowing on lung X-ray
- phosphate binding in GI tract
  - aching muscles
  - rickets
  - osteoporosis
- skin reactions (from aluminum antiperspirants)
  - milaria (acute inflammation of sweat glands)
- encephalopathy
- senile dementia (Alzheimer's disease)
- nephritis
- hepatic dysfunction
- gastric distress
  - GI inflammation

flatulence and acid eructation (belching)
colitis
hyperactivity in children
psychosis in children

If you work in certain industries, you will have a greater risk of exposure to aluminum poisoning than other people. Obviously, avoid working in these industries if it's at all possible. Here are the occupations most liable to offer aluminum intoxication:

manufacturing of aluminum abrasives
treating bauxite ore to obtain alumina
production of aluminum sulfate (alum) from bauxite ore
manufacturing of aluminum products
aluminum alloy manufacturing
paper industry
glass industry
textile industry (waterproofing)
use of aluminum abrasives in many industrial operations
manufacturing of aluminum metal powders
synthetic leather manufacturing
aluminum welding
porcelain industry
explosives manufacturing
pyrotechnical devices manufacturing and use

Aluminum may interfere with your normal body functioning at levels lower than were previously assumed. The aluminum industry would probably prefer that this information not be generally disseminated, because it might cause you to avoid buying an aluminum product and to avoid bringing it into your home. The following are common household items comprised of aluminum or its derivatives that potentially could expose you to an excess of the mineral. Symptoms of aluminum poisoning may be brought on by their use, particularly if you have an overactive parathyroid gland as a result of a diet too high in phosphorus and too low in calcium and magnesium.

aluminum cooking vessels
baking powder (aluminum sulfate)—as contained in many of our junk foods today
aluminum-containing antacids

deodorants and antiperspirants
aluminum dust from industrial aluminum manufacturing
building construction materials
household and industrial utensils
insulated cables and wiring
packaging materials (e.g., aluminum foil)
fine aluminum powder used in bronze paint
aluminum cans
drinking water (alum used to kill bacteria)
soil (naturally occurring ores) and dust, particularly from a nearby aluminum refiner
coal burning
plants, including those used as food
beer
milk and milk products (from equipment)
alum used in food processing (pickles, maraschino cherries)
medicinal aluminum compounds used externally to treat dermatitis, wounds, and burns
nasal spray (alum)
toothpaste
Ceramics (made from aluminum oxide clay)
dental amalgams
cigarette filters
tobacco smoke
automotive exhausts
pesticides
animal feed
food, drug, and cosmetic color additives
vanilla powder
table salt and seasonings
bleached flour
American cheese
fumigant residues in foods (aluminum phosphide)
Kaopectate and other medications containing kaolin (aluminum silicate)
feldspar and mica
McIntyre aluminum powder (used in prophylaxis of silicosis)
aluminum silicate paste (arthritis treatment)
sutures with wound-healing coatings containing aluminum
aluminum chelates of polysaccharide-sulfuric acid esters for peptic ulcer treatment
aluminum nicotinate (hypercholesterolemia treatment)

CADMIUM

Smoke a pack of cigarettes and you increase your body burden of cadmium by 4 mcg a day. There is an average of 1.4 mcg in each smoked cigarette, which the body can counteract to some extent through its anti-stress enzymes. But a little of the cadmium poison remains from each cigarette and accumulates, eventually to harden your arteries, destroy a few thousand brain cells, produce lots of skin wrinkles, and do other untold damage to shorten your life. All smokers have twice the body levels of total cadmium as nonsmokers. The latest research from Amsterdam has shown that cadmium can increase the basement membrane thickness of tiny blood vessels and capillaries, particularly in sensitive tissues such as the uterus. Thickening of the basement membrane tends to impair circulation. If you impair circulation in the uterus, a pregnant woman will be starving her fetus while it is forming. The result will be either low birth weight or prematurity, or even deformities in the newborn as a direct result of smoking while pregnant. The evidence is in. Newborn tragedies are occurring regularly in our society today. No one should smoke while pregnant unless the mother is prepared to spend a lifetime in remorse, helping a deformed child through life.

The usual signs and symptoms of cadmium toxicity are the following:

fatigue
hypertension (possibly related to increased concentration of cadmium in that part of the kidneys responsible for their main function)
iron-deficiency anemia
emphysema
Itai-Itai (bone softening in women over forty years of age who had at least one child and suffer with dietary deficiencies such as those in the third world)
slight liver damage
anosmia (loss of sense of smell)
yellow coloring of teeth
reduced birth weight in newborns
renal colic (with passage of calculi)
nephrocalcinosis
hypercalciuria (excess calcium in the urine)
pain in lower back and legs
pain in sternum
"milkman's syndrome" (lines of pseudofracture in scapula, femur, ileum)
hypophosphatemia

possible rheumatoid arthritis
decreased production of active vitamin D
decreased pulmonary function
proteinuria, glucosuria, and aminoaciduria (following nine years of inhaling cadmium oxide)
possible prostatic cancer (in workers exposed to cadmium oxide)
possible carcinogenesis
increased mortality

You may not smoke, but if you are employed where cadmium exposure is excessive, the mineral will leach into your body system whether ingested, inhaled, or absorbed through the skin. The following occupational exposures of cadmium are relatively high:

nickel-cadmium battery manufacturing
zinc or polymetallic ore smelting
paint manufacture using cadmium pigments
painting with cadmium pigments
jewelry making
cadmium alloy manufacturing
ceramic making using cadmium
electroplating metals with cadmium
process engraving
cadmium vapor lamp manufacturing
rustproof tools, marine hardware, etc.
soldering
tetraethyl lead manufacturing (uses diethyl cadmium)
fungicide manufacturing

Environmental cadmium has increased strictly as a result of greater industrial use. Once it is absorbed, you can hardly get rid of it, and the metal's toxicity is significant. It makes one feel despair to realize how people destroy themselves with cadmium pollutants. Chelation therapy holds out the primary hope for treatment of cadmium toxicity. The sources of cadmium listed here are common items to be avoided:

drinking water
soft water, causing uptake of cadmium from galvanized pipe
soft drinks from vending machines with cadmium piping
refined wheat flour (increased cadmium: zinc ratio)
evaporated milk
many processed foods

oysters, kidney, liver
rice (irrigated by cadmium-contaminated water)
soil
cigarette smoke
tobacco
superphosphate fertilizers
cadmium alloys (e.g., dental prosthetics)
ceramics
paint pigments
electroplating
cadmium vapor lamps
tools rustproofed with cadmium
marine hardware rustproofed with cadmium
welding metal
solders
silver polish
polyvinyl plastics
fungicides
pesticides
sewage sludge and effluents
copper refineries
dust in urban streets, homes, businesses, and schools
rubber carpet backing
wear of rubber tires
burning of motor oil
plastic tapes
black polyethylene
black rubber

CANCER PREVENTION AFTER EXPOSURE TO LEAD

"Lead concentrations in the human body tend to encourage the incidence of cancer," wrote W. Blumen, M.D., Chief of Oncology at a medical clinic in Zurich, Switzerland, in a 1976 medical journal.[1]

During the years just prior to 1959, forty-seven adult residents of a local Zurich community who had complained of symptoms associated with lead toxicity were treated with EDTA chelation therapy. Living adjacent to a major traffic artery, they had inhaled high quantities of lead present in auto exhaust. They experienced headaches, fatigue, stomachaches, intestinal ailments, acute depression, and other health problems. All these symptoms were found in the forty-seven residents with particular frequency, and they also showed lead deposits lining their teeth at the gums (plumbism). An inordinate amount of delta-aminolaevulinic acid appeared in their urine (an early indication of lead poisoning). As a result, their lead intoxication was treated with EDTA chelation therapy, under the auspices of the Zurich Board of Health, and eliminated, allowing them to go on living relatively healthy lives.

In 1972 Dr. Blumen and his co-investigators were studying the incidence of cancer among 232 adults living along the same road. The researchers observed that 11 percent of the 232 residents had died of cancer between 1959 and 1970. "This percentage was nine times higher than that observed in a traffic-free region of the same community," states Dr. Blumen.

Of 25 individuals residing exactly adjacent to this well-traveled traffic artery who died of cancer, 22 had lived there at least ten years prior to their deaths. The diagnoses made at the hospital for all these cases were five lung cancers, four colon cancers, three stomach cancers, two breast cancers, three ovarian cancers, one pancreatic cancer, and seven cancers at various body locations where such occurrences are rare.

Up to 1970, only one of the 47 residents treated previously with EDTA chelation therapy had incurred cancer and died. By contrast, of the other 185 residents—those who did not receive the EDTA treatment—24 became victims of the disease. This translates to a cancer percentage increase of six times over the chelated people. Cancer mortality in all age groups was thus much higher in the untreated individuals, or—to put it another way—600 percent less cancer appeared in people receiving EDTA intravenous infusions for a reason not specifically related to cancer.

Can chelation therapy prevent the occurrence of cancer from lead and other pollutants in our industrialized society? Is chelation therapy a partial answer in the treatment of some miscellaneous malignancies? To both questions, my response must be a qualified "yes."

The growth of transplanted T8-Guerin-carcinomas (tumors caused by virus) in rats has been prevented through intravenous injections of EDTA. This preventive effect is attributed to the fact that metal ions that are indispensable for the growth of tumors were eliminated by the chelation technique. This has become standard procedure among some of those using metabolic cancer therapies.

It turns out that almost every chemotherapy procedure used in oncology today involves chelation. Nearly all the anticancer drugs are chelating agents; most antibiotics are chelators; all the heavy metal detoxicants are chelators; many pain relievers are chelators.

The *German Journal of Oncology* has reported that enzymes will strip the outer coat of protein away from the cancer cell. This is one reason why enzyme supplements in high doses are frequently prescribed today by physicians interested in alternative, less toxic cancer therapies. However, there is a deeper cleavage of protein that needs to be removed by chelation before the cancer cell will be fully exposed to the body's immune system. For this purpose, a chelating agent is presently undergoing investigation at the Downstate Medical Center of the State University of New York. The chelator, penicillamine, is being experimented with for use in certain types of cancer.

Dr. Blumen's published report described how the removal of lead caused a decided decrease in the incidence of cancer, but it may not have been lead detoxification at all that saved them. It could be simply that their improved circulation was responsible. Cancer grows under anaerobic conditions—in areas where there is an impaired oxygen availability. Where chelation allowed the bloodstream to surge normally and carry its full complement of oxygen, cancer cells probably could not take hold in the tissues or organs of these people.

On the other hand, cancer specialists acknowledge that cancer usually is present in your body for a minimum of seven years before it gets noticed as a lump or bump large enough to be spotted on an X-ray film by a radiologist or on the body surface by you. Since lead and other heavy metals chronically depress the function of your immune system, it's well established that getting the lead out of the body occasionally may be a pretty good plan if you want to keep your immune system working at maximum. Doing something to enhance the immune system, such as taking chelation treatment, will stimulate it to attack the numbers of wandering cancer cells that are believed to be present in most of us from time to time. If your immune system is in good shape, your body destroys these early cancers.

THE WORKING OF YOUR IMMUNE SYSTEM

The immune system is your body's major "department of defense" in the battle against cancer or infection caused by invading carcinogenic antigens or microorganisms. Maintenance of the system is crucial to any body's well-being. Without immunity, death can result. Unfortunately, as we age our immune system begins to fail and we become more susceptible to cancer and infection. If we could delay this problem of "immunosenescence," experts on aging believe we could extend our life expectancies.

The human immune system is divided into two components, humoral and cellular. The humoral component involves the interaction between the invading microorganisms, or antigens, and your body's protective antibodies, primarily gamma globulin molecules found in blood plasma. Cellular immunity involves the interaction between carcinogenic antigens and certain specialized cells in the blood and lymphoid tissue called T-lymphocytes.

Both immune components are specific; each antibody or lymphocyte responds only to specific antigens that are "recognized" and "remembered" from previous exposure.

The humoral immune process works by manufacturing antibodies to combat invading microorganisms containing antigens, including viruses, bacteria, and fungi. Antigens stimulate the production of antibodies, or gamma globulin (a plasma protein) molecules. Each antibody has a highly specific affinity for the particular antigen that stimulates its production.

Antibodies are produced primarily by the B-lymphocyte cells located in the lymphoid tissue. Such tissues are found in lymph nodes, the spleen, the gastrointestinal tract, and bone marrow.

Once lymphocytes and their mature descendants, the plasma cells, begin to turn out a specific antibody, they continue to do so for varying periods, sometimes for all of your life, whenever the initiating antigen reinvades your body.

Antibodies combine with the antigen that stimulated their production. Once this reaction has been set into operation, you are said to be immune to that particular antigen.

The first interaction between antigen and antibody is known as the primary immune response. This is characterized by a slow onset of antibody formation with only limited numbers of antibodies produced, and creation of a large number of "memory" cells capable of responding to the same antigen in the future.

A secondary immune response will occur every time the same antigen is reintroduced to your body. This response is characterized by rapid activation and differentiation of the "memory" cells, and creation of large amounts of antibody.

Humoral immunity can be active or passive. In active immunization, antibody production is stimulated by the administration of antigen or by exposure to naturally occurring antigens. In passive immunization, preformed antibodies (gamma globulin) from another person or animal are administered to you. These donated antibodies provide short-lived and potent immunity for you.

An example of active immunity would be measles—measle virus enters a person's body and initiates antibody production against the disease. A person who has had measles once is usually immune for life. Active immunity can also be initiated deliberately, as with vaccination. The patient is injected with a small amount of modified antigen, which is harmless but can stimulate antibody production and prime the defense system for subsequent exposures. For example, pertussis vaccine creates a long-acting protection against whooping cough.

An example of passive immunity is that imparted by injection of immune globulin for prevention of hepatitis A. If the injected patient encounters the disease, he already has a supply of preformed antibodies to fight it.

Cellular immunity and delayed hypersensitivity are functions of the T-lymphocyte cells, also found in lymphoid tissue. The cellular immune process is responsible for rejection of transplanted organs, response to slowly developing bacterial disease, such as tuberculosis, and action against cancer cells.

The cellular immune response is initiated when a T-lymphocyte cell encounters a specific cancer or other antigen and releases numerous chemical factors that kill the invader. Several other more complicated steps are involved in cellular immunity, which is a much slower-acting process than humoral events.

Unfortunately, due to aging and for a variety of other reasons, including nutritional deficiencies and the overwhelming pollution in our environment, the cellular immune response does not get initiated. It's not entirely known why, but the T-lymphocytes don't recognize that cancer cells are the enemy. Cancer fails to be destroyed and begins its slow, insidious growth over a period of several years while you continue to be unaware of it. If you happened to have taken chelation therapy for lead poisoning or for any other reason, AAMP physicians believe, the EDTA or other chelating substance will be able to strip the protein coat around any tumor cell. It's the protein coating that prevents T-cells from attacking cancer as the enemy, because it cannot identify it. When the chelator has done its job of stripping the cancer cell of its protection, your immune system identifies it. No longer can the cancer cell sneak into a tissue like enemy aircraft under your body's radar system. Chelation has helped sound the warning two ways: First, it removed the excess lead burden we all have today so that the immune system was vitalized; second, it brought the cancer cells out of hiding behind their protein coating so that the T-lymphocytes could spot the tiny cancer growth and shoot it down before its full invasion.

Besides being a possible prevention method against cancer, intravenous chelation has significant potential for use in people with active cancers who are undergoing chemotherapy. "Aggressive chemotherapy depresses the body's immune system, leaving the patient open to infection," points out

Bernard Pirofsky, M.D., Professor of Medicine and head of the University of Oregon's Division of Immunology/Allergy and Rheumatology in Portland, in an interview. "Cancer, burn transplantation and immunological disease patients all have weakened immune systems." There are thousands of cancer patients undergoing such aggressive chemotherapy in the United States at any one time who could benefit from anything that would enhance their immune system. One possibility is intravenous infusions with one or a combination of chelating agents, given with supplemental nutrients such as vitamins A and C, zinc, selenium, and other nutrients discussed in Chapter 7. They also enhance the immune system.

Among the worst immunological stressors is being burned over a part of your body. "Severe burns suppress the efficiency of the body's immune system through the loss of antibodies," Dr. Pirofsky explained. Approximately 70,000 Americans are severely burned each year. Also, according to the immunologist, "Therapy required for transplant operations alters the function of the patient's immune system. Patients need a full range of antibody protection prior to, during, and after the operation. . . ." Kidney transplant patients alone number approximately 5000 yearly. They could be the beneficiaries of the immunological enhancement attributes of chelation nutritional therapy.

A number of diseases are characterized by abnormal immune function. These include relatively common conditions such as rheumatoid arthritis, multiple sclerosis, lupus erythematosus, and various other immune-system-related disorders, including aging. "Some of the therapies used [in these conditions] depress the immune system, increasing the possibility of infectious disease complications," explained Dr. Pirofsky, who is also a rheumatologist (arthritis specialist).

Some recent research has shown that acute rheumatoid arthritis responds beautifully to penicillamine, a chelating agent most specific to copper. Does this mean that the injection of penicillamine will quiet down the inflammation of rheumatoid arthritis because the chelator has removed copper from the arthritis victim's body? While this premise sounds reasonable for the diminution of arthritic symptoms, the treatment doesn't act physiologically in just that way. The chelation answer is that if you are chelated for relieving rheumatoid arthritis, your level of a particular mineral such as copper will be changed in the bloodstream. The mineral's availability to certain cell processes is altered by the chelating agent. The combination of penicillamine and copper creates a potent antiinflammatory element in your blood, working directly on the excited tissues. Your painful knees, feet, elbows, and hands will suddenly stop hurting and become more flexible because a redistribution of a bioavailable form of copper has taken place. There may even be some lopsided logic to the superstition that wearing a copper bracelet wards off arthritis.

When biochemists tie the chelating substance aspirin to copper, they create a new molecule of copper aspirinate. This new experimental molecule is legal to use and becomes another example of the synergistic effect furnished by two antiinflammatory substances coming together. Aspirinated copper will provide a person in pain with about ten times the antiinflammatory potency of cortisone, without cortisone's side effects. Copper aspirinate is carefully prescribed by knowledgeable physicians who regularly practice these metabolic methods. They make copper available in this manner for a particular part of the cell metabolism, where it produces a powerful antiinflammatory action.

For a reason similar to this movement of minerals for cell metabolism, people suffering with severe leg cramps as a result of hardening of the arteries will often find marked and surprisingly rapid relief after taking only a few infusions of chelation therapy. The ethylene diamine tetraacetic acid injections may have not yet significantly improved their arteriosclerosis but it may provide results because it can help make the mineral zinc more available to the body's nicotinamide adenine dinucleotide (NAD) pathways. NAD affects skeletal muscle cell metabolism, and the legs move through the action of skeletal muscles. Everyone should ideally have approximately eight parts of zinc available to every one part of copper present in his body. If he does not, metabolism could suffer. The muscles may complain with cramping, especially the calves, which are the longest and strongest muscles in the body. Zinc EDTA may increase availability of zinc to calf muscles hungering for zinc. These muscles finally receive the missing mineral for which they have been screaming with pain and stop their cramping, although it may also be a magnesium or calcium need that must be met. Frequently oral magnesium supplements help this condition as well.

If a physician was cognizant of trace elements and chelation therapy, he could give his patient suffering with leg cramps immediate and usually lasting relief using many different approaches. For example, a mild chelating agent such as lactated Ringer's solution with 2000 mg ascorbic acid and 80 meq (milliequivalents) of magnesium chloride could be injected intravenously and repeated two to three times a week for the same number of beneficial treatments (twenty to thirty) usually given with EDTA. This combination of chelators will give nearly the same efficacious circulation response as EDTA. If a physician believes the erroneous information about nephrotoxicity response with EDTA, let him give his patient a break and at least prescribe by mouth or inject one or a combination of the milder and newer chelators described in Chapter 7.

Ringer's lactate with ascorbic acid will lower the ionizable calcium in any monitored patient to approximately one-half normal. Simply testing his patient's ionizable serum calcium before he begins the bottle of lactated Ringer's solution with vitamin C into a vein and testing again during the infusion will let any physician discover that the patient's ionizable serum calcium has been dramatically reduced. The mild chelate pulls ionic calcium out of the person's blood, to be replaced later when the parathyroid gland moves into action.Parathormone, seeking a renewed source of calcium, will put back into solution any calcium it can find. Some of it will come from the patient's metastatic calcium which has been lodged in the cells of his arterial walls and is impairing these cells' function. The overall chelation treatment effect is to reduce the patient's dysfunction from excess calcium, such as the increased spasm in the arteries, which is making his heart work too hard and decreasing the amount of blood it can pump out with each beat.

The scholarly medical research team of Drs. Peng, Kane, Murphy and Straub at the Veterans Administration Hospital in Little Rock, Arkansas, has confirmed with laboratory experiments on isolated heart cells that calcium-chelating agents, in particular EDTA, restore the high-energy phosphate (ATP) bonds required for energy. They become depressed when the blood supply to the heart is diminished. The doctors report that there is restoration of normal oxidative metabolism in blood-starved cardiac muscle cells after perfusion with calcium-chelating agents; their paper, published in the prestigious *Journal of Molecular and Chemical Cardiology*, tells of the conversion of ADP to ATP for energy in damaged heart muscle cells inhibited by excessive intracellular calcium ions, and states: "this inhibition can be partially reversed by the addition of calcium chelating agents."[2]

OTHER PREVENTIVE BENEFITS FROM CHELATION THERAPY

EDTA infusion may work in a way similarly to papaverine. Papaverine is a nonnarcotic drug with a mild analgesic action. Ingested EDTA reduces pain through inhibition of phosphodiesterase enzyme.[3] Phosphodiesterase enzyme produces an increase in the concentration of cyclic adenosine monophosphate (AMP), an enzyme in your body, resulting in increased breakdown of glycogen to glucose for easier and more effective sugar utilization.

When EDTA medication is directly administered into the vascular compartment by intravenous injection, the infusion produces effective drug levels in the tissue far above those obtained with any oral medication. There are also potential medication effects from the EDTA injections superior to those obtained with known oral vasodilators.[4] In effect, taking a drug for some health reason and having undergone chelation therapy, you'll find the drug works a lot more effectively and a smaller dose will be required.

Intracellular lysosome function of your cells will have been impaired by the accumulation of toxic metals over the years. Chelation therapy cleans from the lysosome an accumulation of wastes that may be impairing its function. The function of the lysosomes in your cells is somewhat like the action of a toilet in your home. The lysosome helps to flush a cell of its waste products and destroy accumulated toxins. When toxic trace metals such as lead, mercury, cadmium[5,6] or excess levels of calcium or zinc[7] block the lysosomal membrane, the function of the lysosome will be blocked, too. Blockage may contribute to the development of many chronic degenerative diseases such as arthritis, multiple sclerosis, lateral sclerosis, parkinsonism, and more.[8] Removal of these heavy metals will allow the lysosome to detoxify more efficiently—even to improving myocarditis due to lead poisoning.[9]

Any toilet that becomes stopped up begins to back up. Like a septic system, lysosomal membranes that are blocked cannot get rid of toxic materials. Because EDTA is an amino acid, it does not just float in the bloodstream, but also profuses through the tissues, the capillary bed, and the tissue fluids, in that way pulling out toxic metals from the sixty trillion cells of your body. When EDTA gets near a cellular membrane it binds with any toxic divalent or trivalent mineral, which is a heavy metal such as lead, tin, mercury, aluminum, and others that are impairing membrane function and contributing to free-radical damage and lipid peroxidation of these important cellular membranes. The chelating agent pulls out the toxic mineral. Other toxins are then floated from the cell by osmosis because now there is a greater concentration of toxins inside the cell than outside, and those toxins eventually will be excreted from the kidneys.

The incidence of aging is slowed down by EDTA infusion. One of the reasons this may occur is that EDTA alters tryptophan metabolism.[10] Tryptophan is an amino acid component of proteins. When the body content of tryptophan is lowered, your life-span extends. This is an antiaging effect of EDTA chelation that may warrant further investigation by gerontologists.

In the book *Prolongevity* (Alfred A. Knopf, 1976), author Albert Rosenfeld suggests that, in time, we can slow the rate of aging, alter the biological clock that may be built into our cells, give our lives a kind of biological daylight saving time, with more sunlight at the end. Science writer Rosenfeld reviewed and explained a great deal of aging research, but the ultimate technical marvel for slowing aging has been with us for thirty years, and he overlooked it. EDTA chelation therapy is a gerontologist's rose garden in full bloom.

The flow of fresh blood throughout the capillary bed is markedly improved by EDTA chelation as a result of the thinning of the blood, as well as reduced blood fat levels.[11] Lowering blood fats contributes to decreasing the rouleaux formation by the red blood cells. Rouleaux is the stacking or clumping together of red blood cells just like coins rolled into a wrapper. Chelation therapy decreases this red cell aggregation (rouleaux), which helps to speed blood circulation. Also, chelation therapy is of further direct benefit to the circulation because it reduces the rigidity of red blood cell membranes.

The improved pouring in of blood to the capillary bed throughout the body, as well as better peripheral blood flow in the extremities, also comes from the decreased resistance of the blood vessels.[12] Furthermore, the resilence, elasticity, and compliance of other vascular structures are improved by the decalcification of elastic tissues, with associated decrease in cross-linkages in these tissues.[13-15] An improved capillary bed occurs additionally from what Gus Schreiber, M.D., of Dallas, Texas, described in an unpublished paper[16]—the decrease in basement membrane thickening, particularly in diabetics. Dr. Schreiber performed electron microscope studies of thigh muscle biopsies before and after these biopsied patients underwent chelation therapy.

Rheumatoid arthritic symptoms are prevented by collagen tissues being aided in their formation by the high parathormone levels. Diseases around joints take place by irregular collagen development. Higher parathormone levels, you may recall, are produced in response to the abnormally low levels of calcium in the circulating blood due to the binding of serum calcium by EDTA and its subsequent elimination through the kidneys. Rheumatoid arthritis and other collagen diseases are involved with degradation of mucopolysaccharide and protein connective tissues. Parathormone alters the turnover of these tissues and thus affects rheumatoid arthritis pathology.[17,18]

By altering your membrane calcium components, EDTA chelation increases red blood cell membrane flexibility.[16,19] Greater flexibility allows the red blood cells to fold more easily to conform to the tiniest of capillaries, which may be even smaller than the cells themselves. As a result, sickle cell anemia has been reported in the journal *Hospital Practice* to improve from EDTA injections because of the reduced cell wall rigidity.[20] Less rigidity also permits potassium to enter into the cells more readily.[21,22]

Administration of EDTA chelation lowers your blood pressure levels. This arises from an increased excretion of cadmium from kidney tissue cells,[23] the decreased peripheral resistance[12] through greater resiliency of blood vessels after removal of calcium[13-15] and associated decreased vascular spasm. Moreover, lowering of your high blood pressure takes place from increased serum magnesium, as previously mentioned.[24]

EDTA infusions protect you against cirrhosis or impairment of liver function in high-fat diets.[19] The treatment improves lipid metabolism[25,26] and enhances glucose metabolism in diabetic patients.[27,28]

EDTA chelation alters cross-linking of elastin, the major connective tissue protein of elastic structures.[13,14] In large blood vessels and elsewhere, elastin is a yellow, elastic fibrous mucoprotein. This elastin alteration happening in the aorta has profound consequences, since the cross-linking of macromolecules by free radicals and calcium is now thought to be a basic aging mechanism.[29] Furthermore, an increased stiffening of the aorta, occurring in all of us by age fifty, markedly increases the work load of the heart as it pushes the blood out each time against this stiffening resistance. It's a little like someone pushing against you from the other side of a door you want to go through.

EDTA chelation lowers your serum phosphorus levels, which affect phospholipids advantageously.[30]

By means of its binding with zinc, EDTA chelation assures you of more adequate levels of crucial unstable zinc.[31] Three components—the lactate dehydrogenase enzyme (LDH) of skeletal muscle and nicotinamide adenine dinucleotide enzyme (NAD) involved in degrading lactate and priming the change of glycogen into glucose, and any damaged cardiac muscle—are improved in their "demand-adaptation" by this change in zinc status.[32]

After having no viable approach for relief of significant vascular illnesses, the medical profession now has available intravenous EDTA chelation. It is a therapy highly acceptable by patients. It brings not only physiological benefit but psychological comfort as well. Psychological relief is obviously going to occur, because chelation removes people's anxiety about dying or losing a limb or an organ or facing dangerous surgery, or having no possible help for their condition. Lowering anxiety levels is already proven[33-35] to increase circulation due to vasodilation, similar to that taught in hand-warming techniques used in biofeedback training.

Insulin requirements are reduced temporarily and good control may be obtained in some diabetics simply by extending the therapeutic action through repeated EDTA infusions at appropriate monthly or bimonthly intervals.[27,31]

Diabetes mellitus has a well-known relationship to atherosclerosis which is of great importance in chelation therapy. In a study of a large group of patients with atherosclerosis who didn't show any of the typical clinical features of diabetes, 56 percent were found to have latent disturbances of carbohydrate and sugar metabolism when adequately tested with a glucose tolerance test and insulin level determinations. The insulin disturbances were highest among those with advanced coronary artery heart disease.

In Dr. Norman E. Clarke's series of many thousands of patients with advanced forms of occlusive vascular disease, those with diabetes mellitus responded most dramatically to chelation therapy. Dr. Clarke, when he spoke before the California Medical Association ad hoc committee in March 1976 (see Chapter 5), said that there has been a great reduction in insulin requirements among his patients. He checks their insulin needs frequently and lowers them gradually to avoid a severe insulin crisis. Other chelating physicians have repeated this same observation, even reporting improvement in some brittle, previously uncontrollable severe juvenile diabetics.

The procedure followed by AAMP physicians is to first put diabetics on a special chromium supplement known as glucose tolerance factor. They also prescribe a high-fiber, low-fat diet with exercise. Then chelation therapy is given when and where needed to avoid and control the vascular disease that all diabetics are known to get worse and faster than anyone else.

Blood fat, as I said, is decreased by up to 33 percent.[7,32,33] Therefore, a person with an elevated level of cholesterol in his blood, present for a long time, is more easily managed after chelation by his physician using the antihyperlipidemic supplements employed by nutritional physicians, including high fiber, vitamin E, selenium, iodine, magnesium, niacin, etc.

Elasticity is restored to any of your rigid, nonstretchy arterial walls. The excess cholesterol in the atheromatous plaque is the product of impaired metabolic processes and prior degenerative reactions within the arterial wall. You probably have thickened fibrocollagenous tissue producing localized artery wall enlargement with early obstruction at points of excess hemodynamic stress, but the EDTA treatment program helps to correct this condition.

Medical investigation of human aortas during autopsies has demonstrated that people have alteration in their medial elastic tissue with calcium content increased prior to the appearance of any atherosclerotic plaque. An abundance of mucopolysaccharides in the arterial ground substance serves as a cement between and within the cells. Young men who have died from accidents possess many enzymes such as proteolyses in their arteries that permit restoration of arterial wall injury. In older arteries this restoration process is much reduced, but EDTA chelation allows for its return and rejuvenation, through restoration of previously depressed arterial enzyme function.

EDTA chelation offers you protection against the precipitation of lipoproteins that are produced by heparin. Heparin is the complex anticoagulant principle that prevents platelet agglutination and thrombus formation. Precipitation that is injurious to your health happens in the presence of divalent metals in the bloodstream, such as calcium.[36,37] Heparin is commonly used during acute heart attacks as crisis care, sometimes without the emergency physician recognizing the danger of furthering the local lack of oxygenated blood, which is called ischemia, due to aggravating the mechanical obstruction to the blood supply of the heart.[38] By binding calcium and other divalent cations, EDTA chelation can help to prevent this heparin lipid precipitation and improve cardiac arrhythmias as well.[21,22]

Another important benefit of the internal EDTA chelation mechanism is the dissolving of small thrombi while the chelating agent is being injected. The lowering of your serum calcium produced by the infusion provides less likelihood of abnormal clotting.

To prolong these benefits for you, particularly if you have just suffered a stroke or heart attack, the infusions could be given more slowly such as over a twelve-hour period, which would be ideal for any patient in the hospital after a heart attack. If you don't mind spending so long a time in your doctor's office, your might ask him about being infused for a longer than four-hour individual treatment, particularly if you are very weak, or find the more rapid infusion uncomfortable in spite of additional Xylocaine, which is all most people need to be completely comfortable.

As has been mentioned, often patients voluntarily return to their physicians after one or two years for a short series of chelation infusions to prevent any loss of the remarkable benefits they obtained from their initial series of treatments. Chelated people do understand that the calcification and atherosclerotic process is an ongoing process of degeneration of arterial walls. Our bodies constantly are fighting with this deteriorating process. Knowing this, some patients feel so gratified with their health improvement, having previously suffered from severe illness, that they continue to add to their total number of chelation treatments by getting twelve to twenty-four additional treatments with each passing year. This is good business for the chelators and, more important, life-preserving maintenance for the patients.

There does not appear to be any limit to how many repeat infusions can be taken.[25,39,40] Individuals have received over 500 infusions in a ten-year period; and where the previous medical history had shown several strokes or myocardial infarctions before chelation with EDTA, they frequently have had no further vascular events during this extended treatment time. Because they were considered high-risk patients before they started chelation treatments, cardiovascular victims are *not* taking any chances.

In summary, more and more our Western industrialized society is going to show signs of heavy metal poisoning such as from mercury and lead, particularly as automobiles and other pollution sources overtake us. Stricter controls on auto emissions did not go into effect until the 1980 model year. Now the Reagan Administration supports loosening of those controls. The U.S. Congress delayed cracking down on the car fumes four times in knuckling under to oil and automobile financial interests in the years prior to 1980. It is likely that Congress has contributed to the impaired health of many U.S. citizens by its emission controls deferment. By his bowing once again to vested commercial interests, President Reagan will be adding to the numbers of Americans suffering with toxic-metal-induced degenerative diseases.

When asked at one point why pure air should be protected at the expense of those who favor building power plants, former Secretary of State Edmund S. Muskie, who was then the United States Senator from Maine and who had sponsored a bill to amend the clean air act that was favored by environmentalists, was quoted in the June 10, 1977, issue of the *New York Times* as saying: "Maybe you want to feel you are part of the Earth that is close to what God created. That is what we are trying to preserve."

Inasmuch as this is a book about the chelation answer to hardening of the arteries, something had to be said about heavy metal toxicity. However, I have to admit that a great deal more remains to be explained. The relationship of chelation therapy to mineral metabolism is dramatic. A smattering of information is supplied in this chapter, and I expect that it has opened your eyes to the potential poisons around you in food, air, water, your home, your workplace, and areas of leisure. What should you do to preserve our environment whose despoilment has such a devastating effect on your health? First, take holistic health measures described on these pages; second, put political pressing on the politicians representing you. They are the ones who may finally legislate whether you die prematurely or live to a healthy old age.

CHAPTER NINE

Coronary Artery Bypass in Contrast with Chelation Therapy

What's it like to have a heart attack?

"It's quite horrible," say Allen Farcas of Baltimore, who has survived two of them. "Besides being debilitating, it is demoralizing. Being admitted into the intensive care unit is a life-or-death matter, and you know your life hangs in the balance if you're conscious.

"The actual attack is an absolute torture, the pain penetrates the heart so intensely. It radiates over the chest and deep within as if an arrow is stuck in your breastbone. Having that pain once, you never want it again. That's why I don't understand why anyone settles for patchwork treatment of his heart disease and doesn't radically change from abusive habits that have brought his body to this weakened state in the first place," Farcas added.

"It's true that it took a second escape from death before I became frightened enough to stop smoking, went on a sensible 'stone-age' diet, and made a daily session of exercise part of my lifestyle. I also investigated alternatives to coronary artery bypass surgery," Farcas said, "and decided that this life-or-death operation is not something for me."

Bypass surgery, which has briefly been described in other chapters, is a shunting operation. A vascular prosthesis is used between the aorta and the femoral artery to relieve obstruction of the lower abdominal or at its bifurcation or at the proximal iliac branches (the aortoiliac bypass); or the bypass may be another operation to shunt blood through vein grafts or other conduits from the aorta to branches of the coronary arteries, in order to increase the flow beyond the local obstruction (the coronary bypass); or it could take the form of a third operation in which a vascular prosthesis is used to bypass an obstruction in the femoral artery, where the graft may be a synthetic material such as Dacron, or autologous tissue such as bovine carotid artery (the femoropopliteal bypass).

The coronary bypass is the most popular chest surgery done today. It takes tremendous skill, much training, and a large investment in the surgeon's education, along with some highly sophisticated and costly equipment supplied by the hospital involved. The whole procedure must go quickly and cleanly, with great efficiency of motion and time. Every health professional participating works as part of the surgical team. First you receive an angiogram; then a shave and cleaning of the chest. The anesthesiologist arrives to put you to sleep, and then the operation begins.

Suppose you are a person like Farcas and become the victim of a heart attack—the result of organic disease affecting your coronary arteries.[1] The coronary arteries supply blood to the heart. They have narrow openings, the widest being perhaps the circumference of a small pea. Atherosclerosis is apt to attach its fatty plaques to the intima of these miniature but vital blood vessels. Calcium penetrates their cell walls, causing sporadic spasm. An effect of coronary artery plaque formation and arterial wall spasm will likely be an insufficient supply of blood to a patch of muscle in the heart. The resulting oxygen starvation causes pain that is not felt directly *in* the heart, but is "referred" by afferent and efferent parasympathetic nerve fibers.

The *afferent fibers* transmit impulses to the central nervous system from discrete cardiac receptor endings of various types and from terminal networks that are plentiful in such reflexogenous zones as the endocardium (the innermost tunic of the heart).

The *efferent fibers* carry impulses that are modified reflexly by afferent impulses from the heart and great blood vessels. Efferents are under the overall control of the higher centers in the brain, the hypothalamus (a gland in the brain), and the brain stem.

When impulses pass along these afferent and efferent nerve fibers referred pain permeates muscles in the chest, neck, and arms. This is the pain of angina pectoris (from the Greek: *angina* means "pain" and *pectoris* means "chest muscles"). Several million Americans suffer these same kinds of chest pains, usually from coronary heart disease.

Additionally, every minute of every day in this country somebody has a heart attack—what is known medically as coronary thrombosis (or coronary occlusion or myocardial infarction).

In heart attack there is a difference in the degree of damage to the vital pumping organ. For instance, the coronary arteries may become severely narrowed (coronary occlusion) and produce an attack of mild intensity. On the other hand, a blood clot may completely block the lumen (inner channel) of the coronary arteries (coronary thrombosis) at the point of severe narrowing and cause instantaneous death.

About two out of five heart attacks *do* result in sudden death. Of the 60 percent of patients who survive their initial cardiac occurrence, about 85 percent may go to live an "average" life expectancy of ten more years. These cardiac patients become potentially chronically ill people in danger of sustaining another heart attack unless they live the healthiest of lifestyles.

Although heart attack symptoms may make themselves known all at once, the atherosclerotic condition underlying cardiac disease has been developing for years in your arteries. Consequently, there are many warning signals of cardiac involvement to look for, especially in the dangerous fourth, fifth, and six decades of life, when heart attack is known to occur most frequently (especially among males).

TWELVE SEEMINGLY UNCONNECTED SIGNALS OF CARDIAC INVOLVEMENT

Over time, the various multiple causes and conditions of disease that involve the heart and build up gradually may prepare your natural defense mechanism to assist itself. This will be done by *collateral circulation,* which is a detour of the blood around the blocked artery by way of other blood vessels. Your body is a self-repairing servomechanism, but it responds well to any outside healing asistance.

You can help your own defense mechanism if you know the signals of something going wrong inside. While your body is traveling its road to self-recovery from stress or possible illness, there are at least twelve seemingly unconnected symptoms to watch for that are warnings of cardiac involvement.[2] They are:

1) A persistent sensation of coldness, numbness, tingling, or burning in the toes or feet.
2) A feeling of weakness that comes on occasionally, either generally or to one limb. It may be the sort of weakness that causes you to feel much labored when climbing stairs.
3) A small ulceration of the skin on the ankle or foot that refuses to heal.

4) A white arc in the iris of the eye.
5) A clouding over the lens of the eye by a cataract.
6) Blurred or darkened vision that comes on sporadically such as in TIA (transient ischemic attack—a short-lived type of local anemia caused by obstruction of the blood supply).
7) Shortness of breath that makes walking quickly an event of surprising exertion. Perhaps you have a need to stop for frequent rest periods.
8) Occasional bouts of dizziness or light-headedness or unusual spells of hearing odd noises.
9) The ears may seem to be playing tricks on you with periods of ringing, sensations of vertigo, or sudden partial deafness.
10) Fainting or blackout for no apparent cause.
11) Memory loss and lapses into confusion may be noticeably frequent and be quite an inconvenience. You might be unable to make decisions or find yourself living in the past much of the time rather than in the here and now.
12) Frequent swelling of the ankles unrelated to injury or other cause. Such swelling may even come on daily and last throughout the day.

Once you have become the victim of a heart attack, a series of events suddenly takes place over which you may have no control. Fear, physical illness, and ignorance about what's happening to you may paralyze your ability to make decisions. All you know is that your chest hurts and breathing is difficult.

So there you are, lying on the floor undergoing pain that feels as if your chest is being struck repeated blows with a sledgehammer. Luckily, you're not dead yet, and by chance your family is there, ready to call someone for help.

The ambulance with red light flashing and siren screaming rushes you to a big sterile-looking building. At the hospital they wheel you directly to the cardiopulmonary laboratory. There, some masked and white-coated doctor cuts into your arm or leg (under a local anesthetic) and pushes a catheter through the vein into your heart. A second physician takes X-ray pictures as dye travels through. The films are developed.

You are waiting on the treatment table when a third doctor joins the first two as they look at the angiograms. "You need a heart operation," says the cardiovascular surgeon among them. "It's the only thing to save you—no other alternative. You know that you've lived badly all these years—smoking and boozing and eating junk food. You owe it to your family to go ahead with this operation. We're going to cut out the evil within you."

Stretched out totally helpless, hardly able even to twitch your nose, you stare at these white-coated figures and the starkness of the surroundings. The X-ray pictures of your chest project their images at you from the light boxes mounted on the wall. You can't speak, but you comprehend. Just as the doctor has surmised from the condition of your coronary arteries, you know that you have lived badly these many years.

"Just lie there," you are told as the health professionals go into the hallway to speak with the little woman.

Your wife stands there wringing her hands and being brave.

"Your husband has had a heart attack," says the heart surgeon. "He shows an abnormal angiogram. Come right in here and we'll show you the pictures. There! See how this artery is blocked and that artery is blocked? Your husband must have surgery immediately! Here is the operating permit to sign; he's not capable of signing."

In shock, the little woman cries and then says, "But an open heart surgery is so serious. Isn't there an alternative treatment?"

"There are no alternatives," says the heart surgeon.

Then she asks, "Well, what if . . . what if he doesn't have the surgery?"

"He'll die! Your husband won't have enough oxygen to his heart tissue to live until morning."

She signs, and you undergo a $25,000 operation.

THE NEW MEDICAL-SURGICAL FAD

Just as there are medical differences of opinion about the benefits and risks of coronary angiography, the same situation exists for that other potentially more dangerous vascular procedure, surgical bypass. Coronary artery surgery, also known as bypass surgery, is supposed to improve coronary circulation. It is a controversial, expensive, and extensive operation that has still not proven its value. Possibly coronary artery surgery has some worth for relieving severe heart pain, but for only a limited number of patients with a narrow range of coronary conditions. Or possibly it might be used for relief of severe heart pain that fails to respond to medical therapy, including chelation therapy. Nevertheless, in 1980—a typical year—110,000 people in more than 500 hospitals underwent coronary bypass—about half the heart operations done in this country. Introduced in the 1960s, it has become the new medical-surgical fad for supposed repair of coronary arteries. The reason? Most of those undergoing surgery are told there is no alternative. Do it or die!

James O. Stallings, M.D., a plastic surgeon of Des Moines, Iowa, told me, "The scar of open heart surgery has become a kind of status symbol." In the minds of upward climbers, he said, a scar on the chest seems to have become a sign of social standing and the ultimate consequence of a life of hard work. It supposedly shows sacrifice to one's job and disregard of one's person for the building of a business.

Increasing numbers of patients without anginal pain or any other coronary symptoms are having clogged arteries bypassed in the hope of delaying a fatal heart attack. Yet there is far too little evidence to support this hope. Patients accept high risks. They take a 12 percent chance of inducing a heart attack due to the surgery—during or immediately after the operation. And a 20 to 30 percent rate of closure of their graft in the next two years also must be considered.

Attilio Maseri, M.D., Professor of Cardiology at the University of London's Royal Postgraduate Medical School, has condemned the readiness of American physicians to have their patients go under the knife. Dr. Maseri told the Chicago Heart Association that three-fourths of all coronary artery bypass operations are unnecessary. The patients could be treated just as well and much more cheaply with new drugs such as the calcium channel blockers, he said on February 5, 1981. The calcium channel blockers are just as effective as surgery in reducing pain and preventing heart attacks.

There are 75,000 needless open heart surgeries being performed each year, because doctors assume narrowed arteries are the sole cause of chest pain and heart attacks. "It is true that arteries narrowed by fatty deposits cause some heart attacks," Dr. Maseri said, "but there is also evidence of another cause—spasms of the arteries. These spasms, or convulsive contractions, can occur in either narrowed arteries or in arteries free of fatty deposits, and they can be treated by the new drugs."[3] EDTA chelation therapy is the best and most proven compound among this new class of calcium antagonistic medications just coming onto the market.

The explosive growth in the last ten years of open heart surgical facilities has raised serious questions among philosophers, lawyers, economists, and other physicians about the adequacy of health care planning and controls over the way doctors are creating a demand for this possibly unnecessary surgery. The underlying reason for the surgery's overuse is because it is profitable. Are the doctors offering reasonable disclosures of the risks and the alternatives? In her book *The Unkindest Cut*, Marcia Millman declares that they are not.[4]

Chelation therapy aside, critics of the bypass procedure say that many individuals may have their angina relieved by proper medication, weight loss, and a cessation of smoking. The National Heart, Lung and Blood Institute reported on March 9, 1977, that many patients with unstable angina responded well to drug treatment and had fewer heart attacks after therapy than those who underwent bypass surgery. Angina accounts for more than 75 percent of the bypass operations performed.

Angina patients accept the operation along with its high risks as their only way out because coronary-artery bypass is almost the only treatment third-party health insurance will pay for. In 1981 Harold Cohen, Ph.D., executive director of Maryland's Health Services Cost Review Commission, told a meeting of these questioning economists, philosophers, lawyers, surgeons, physicians, and administrators that in the real marketplace, the operation is boosting insurance premiums. All of us are paying the tab by higher health insurance costs, even if we don't approve of the procedure. Sponsored by the National Center for Health Care Technology and chaired by Elliot Rapaport, M.D., professor of medicine at the University of California, San Francisco, this review meeting was the second devoted to coronary bypass. The operation is decidedly controversial as to its true long-term value. Attendees wondered aloud whether some nonsurgical measures may be getting overlooked in the rush to the operating room. In the case of any surgery, mused Dr. Cohen, less expensive treatments would get greater play if patients were uninsured and had to form "first opinions about their own money instead of getting second opinions on spending someone else's."[5]

Henry D. McIntosh, M.D., of the Methodist Hospital in Houston, Texas, said at a symposium of the American Heart Association in Miami Beach that bypass surgery should be reserved for patients with *crippling* angina who did not respond to more conservative treatment. "I do not believe that surgery is indicated for the asymptomatic patient," Dr. McIntosh said.[6] His comments are totally supported by the National Institutes of Health in its study, which I mentioned earlier.

"Every time a surgeon does a heart bypass," suggests a health care delivery analyst, "he takes home a luxury sports car." The busy cardiac surgeon can easily earn over half a million dollars a year doing bypass procedures. By 1986, if enthusiasts of the operations have their way, hospitals and doctors will collect $100 billion a year doing just this one type of surgery.[4]

"Yes, it's expensive," admitted heart surgeon W. Gerald Austen, M.D., of Massachusetts General Hospital in Boston. Then he sounded a warning: "We have to be careful of overly expanding its use."

Opponents say many of the doctors are cashing in on a lucrative practice, which in turn encourages too many hospitals to invest in facilities that drive up health costs. Greater numbers of smaller hospitals throughout the country that seek more prestige in their communities are developing the expensive laboratory and surgical facilities and medical teams needed to do coronary bypass surgery. Then, to justify the expenditures, keep beds filled, and maintain an appropriate level of staff skill, the hospitals must do several hundred such operations a year. The result is that they go looking for more patients.

Tremendous financial outlays in surgical facilities are likely to create prejudice in favor of using them. This could represent a powerful vested interest group among physicians who feel threatened by an office procedure such as chelation treatment that avoids the need for up to 90 percent of these open heart surgeries. Such situations involving conflicts of interest do exist.

DEATH PERCENTAGES FOR CORONARY BYPASS ARE REVISED

The key to technical excellence for the bypass procedure, as mentioned previously, is expert medical teamwork and careful anesthesia. The operation takes an average of three hours. During the operation, your circulation is temporarily taken over by a heart-lung machine. Medical specialists shunt your blood to oxygenate it while your heart lies exposed. This is the cardiopulmonary bypass, a method of maintaining extracorporeal circulation by diversion of the blood flow away from the heart; blood is passed through a pump oxygenator (heart-lung machine) and then returned to the arterial side of the circulation.

When describing the heart problem of George Frankel, M.D., in Chapter 1, I mentioned that the National Heart, Lung and Blood Institute's recent conference report indicated that the risk of death following artery bypass surgery is between 1 and 4 percent under the best of circumstances, and between 10 to 17 percent under the worst circumstances, such as when a patient experiences severe congestive heart failure. A cardiovascular surgeon at the Stanford University School of Medicine, D. Craig Miller, M.D., said that 1000 Stanford University patients operated on between 1971 and 1975 for coronary artery disease had an 8.7 percent chance of suffering from a heart attack shortly after surgery and a 2.4 percent chance of dying. During the years 1977 to 1979, the degree of risk lessened to half those death percentages. In that period, a 4.6 percent chance of a heart attack following surgery and a 1.2 percent chance of dying occurred at the Stanford University medical facility.[7]

Your survival after surgical bypass obviously hangs by a thread, although the statistics seem to be getting better. Your recovery depends largely on astute postoperative nursing in a coronary care unit, where potentially fatal complications can be detected and treated.

Of the 346 patients receiving coronary artery bypass surgery at the

University Medical Center in Sacramento over a twenty-two-month period in 1979–80, 50 percent either died or suffered a surgical complication during that time. A class-action lawsuit on behalf of survivors of those patients who died was filed during the summer of 1981 against the University Medical Center (UMC), the University of California Regents, and Robert L. Treasure, M.D. Attorney Gordon F. Bowley brought the action on behalf of Mildred Virginia Carpenter of Sacramento, the widow of Edward E. Carpenter. Mr. Carpenter, a plumber, age 62, died July 11, 1980, a few hours after undergoing coronary bypass surgery performed by Dr. Treasure. The cause of death listed on the patient's death certificate was heart attack. The suit alleges that a booklet, titled *Heart Surgery*, handed to open heart surgery candidates or their families at UMC concealed recently revealed high mortality statistics.

Heart attack was the most common form of surgical complication found by UMC cardiologists in their report. While 17 percent of the UMC coronary bypass patients met death on the operating table, another 33 percent died from their surgical complications.[8]

Ira Lang of Stamford, Connecticut, never came out of a coma following his bypass operation. He died ten days after the procedure, in December 1976, also from complications of his open heart surgery.

COMPLICATIONS THAT RESULT FROM CORONARY ARTERY BYPASS SURGERY

The purpose of bypass surgery is to improve the coronary circulation by graft insertion, but 5 to 25 percent of the grafts become occluded within a year.[9,10] And up to 30 percent reocclude in two years. Whether the grafts become occluded or not, after months or years pain may recur with progression of atherosclerosis, reports Paul L. Tecklenberg, M.D., of Stanford University School of Medicine.[11] Some patients are advised to undergo a second coronary artery operation when angina pain returns.[12] And some patients have been known to undergo this dangerous operation up to five times, because the disease keeps returning. The *patient* usually meekly assumes the blame for the failure of the surgery. He knows that he did not alter his style of living after the surgery, which he thought had made him "safe." Therefore, the surgeon is seldom blamed for those treatment failures, says sociologist Marcia Millman.[4]

Coronary artery bypass operations done as an emergency in patients with myocardial infarction can produce operative death rates as high as 30 percent.[13]

In his Stanford University School of Medicine report, Dr. Miller stated that the degree of risk is affected by where the surgery is performed. At the fifteen most prestigious cardiac surgery hospitals, he said, the death rate following coronary artery bypass ranged between 0.3 and 6.4 percent. He believes the improved outcome of surgery in patients presenting with complications is due to greater experience, better diagnostic procedures, and the development of drugs to stabilize the heart before and during surgery.[7]

Operative death for patients with congestive heart failure may be as high as 23 percent.[14] When the very severe complication known as cardiogenic shock strikes, the coronary artery surgical bypass patient death rate approaches 70 percent.[15]

The mortality connected with coronary artery bypass surgery certainly makes the opponents of chelation therapy sound downright silly. They dwell on the less than ten deaths over approximately thirty years in which chelation therapy *may* be implicated, out of tens of thousands of patients who have been helped, while no one stops to mention the thousands of deaths directly and unequivocally due to unproven bypass surgery.

A nonfatal complication of surgical bypass includes perioperative infarction, which is a reocclusion around the operative sites where the graft splices were made. This happens in as many as 20 percent of patients.[16,17] Some patients also experience arrhythmias, persistent postoperative bleeding, pulmonary complications, and infection. A major reason for complication with reinfarction to take place among coronary bypassed patients is that the same conditions exist in their bodies after the surgery as prevailed before. Because traditionally practicing physicians remain ignorant about the value of orthomolecular medicine, these patients are never told that simple supplementation of their diet with safe magnesium, selenium, iodine, vitamins C and E, high fiber, garlic, and other nutrients could offer significant protection. If anything is done, they are probably given drugs, such as aspirin or Inderal. Such drugs may offer more protection than if patients took nothing, but they do not treat the underlying causes of the disease and still allow far too many to die needlessly.

CORONARY BYPASS FAILS TO PROLONG LIFE

According to the Organ Transplant Registry of the National Institute of Health, patients treated with coronary bypass surgery, compared to those treated with drugs, showed no evidence of increased survival or lowered risk of heart attack as a result of the surgery.[18,19]

One randomized trial in forty patients with acute coronary insufficiency involving various coronary vessels compared medical with surgical management; patients who had operations had higher functional capacity four months later, but the rate of death was higher for them.[20]

To date, significant differences in death between the medical management of blocked coronary arteries and surgical management by means of the bypass operation have *not* been reported, except for that March 9, 1977, announcement by the National Heart, Lung and Blood Institute. Michael B. Mock, M.D., the Institute's project officer for its study, said that over four years doctors at eight medical centers across the country tested 288 patients. Of these, 147 received intensive drug therapy and 141 underwent bypass surgery. Mortality in the hospital after treatment was 4.1 percent for the medical group and 5 percent for the surgical patients. In follow-up studies averaging twenty-four months, medical group mortality was 5 percent, against 5.2 percent among surgical patients. In the period of hospitalization, 10 percent of the medically treated group suffered heart attacks, against 18 percent of the bypass patients.[21]

The Medical Letter said, "Although coronary artery surgery may relieve pain in patients with refractory disabling angina, it remains to be established whether coronary bypass operations improve long-term survival."[22]

A previous surgery for heart disease called the Internal Mammary Artery Ligation was abandoned after it was conclusively demonstrated that the operation's relief of angina was all "placebo effect." Some physicians suspect that same placebo effect is partially responsible for the "benefits" of today's popular bypass surgery.

"Many patients do receive pain relief, but the relief is likely to come from the death of the symptomatic area from the heart attack induced at the time of the surgery with subsequent replacement of the area by a scar," Harold W. Harper, M.D., of Los Angeles suggested in our interviews.

"There is also an interruption of the nerve bundle that passes to the affected area of the heart. The pain receptor nerves are severed. As a result, no warning pain is present to signal when the heart is ischemic," said Dr. Harper.

Of ninety-eight patients having coronary artery bypass operations for the relief of recurrent angina pectoris in the Department of Cardiothoracic Surgery, Long Island Jewish–Hillside Medical Center, thirty-eight had their angina pains return after three months. Progressive coronary disease continued in twenty-nine of them (30 percent), and 11 percent of these unfortunates had their grafts close up altogether.[23]

Many patients experiencing failures of their bypass procedures such as those operated at the Long Island Jewish–Hillside Medical Center still deny that they feel severe pain in the chest. Donald S. Kornfeld, M.D., a psychiatrist at Columbia University College of Physicians and Surgeons, believes they have "an investment in feeling satisfied."[5]

Lucien Campeau, M.D., Chief of Cardiology at the Montreal Heart Institute, Quebec, Canada, studied 235 patients angiographically one year after their coronary artery bypass operations. He split them into two groups: those operated on between October 1969 and September 1971, and between October 1971 and September 1973. During those intervals Dr. Campeau found the patient's graft opening had improved from 60 percent to 80 percent. But three years after surgery half of those same patients again studied by Dr. Campeau had no grafts remaining open. Even so, many of the closed-graft patients reported being improved or angina-free.[24]

"This unexpected pain relief may be the result of a 'pain-denial placebo effect,' " Dr. Campeau said. The mere knowledge that they had undergone such a risky and expensive operation may be the cause of the patients' denying the sensation of pain, even if it actually is still present.

Then there are other explanations for no pain. Dr. Campeau said, "These patients may be getting better medical therapy. They're encouraged to stop smoking and lose weight. Many do not resume the same activity after surgery, and many are less anxious because of the operation."

Dr. Campeau cautions that there are serious implications for patients. Like Dr. Harper, Dr. Campeau said, "Not having pain and yet having ischemia might jeopardize patients because they have lost the alarm of pain and might lead too active a life. If they have a positive stress test and still say that they are angina-free," he advises, "they should be watched more closely."

Thus, the quality of life after coronary bypass surgery *seems* to improve when in reality this does not mean the patient has been truly benefited. The patient may only think he is restored to health and is lured to his death through overexercise, suggested Ara G. Tilkian, M.D., of the Palo Alto, California, Veterans Administration Hospital in an interview. Without the chest pain, Dr. Tilkian said, a person can sometimes overexercise himself into new problems.

Exercise that is too severe can cause bypass surgical patients to develop arrhythmias, or irregular heartbeats. "In our studies, one man got a lethal level of arrhythmias," said Dr. Tilkian. The absence of the chest pain removes a natural limiting factor.

WHY CORONARY BYPASS GETS HEAVY REFERRALS

With all the hazards and disadvantages that coronary artery bypass surgery presents to the patient, why do a lot of physicians still recommend this operation? Why are doctors willing to put their patients' lives in jeopardy for so little lasting effect?

Heavy referrals come from some physicians who have been led to believe that not much can be done for angina victims. They try arterial dilating agents first, such as nitroglycerine, papaverine, propanolol (Inderal), and perhaps a hundred other drugs of this type that are medically marketed. Then the doctors may give up on the vasodilators and move to blood thinners such as Coumadin and heparin, and other medications. One of the marks in medicine of an ineffective therapy is that there are so many different drugs to treat the same problem. If drugs were specific and effective, there would be a need for only one. Yet, the public has now been told how marvelous propanolol is for preventing heart attacks, based on a recently discontinued study by NIH (National Institutes of Health).

The November 9, 1981, *Newsweek* reported that the study suggests, according to Peter Frommer, M.D., Acting Director of the Division of Heart and Vascular Diseases, National Heart and Lung Institute, that propanolol could save at least 6500 lives annually. "It was ethically incumbent on us to make it available," Dr. Frommer said.

Unfortunately, it will be years before a study is ever done to compare this propanolol study where the controls received only placebo or nothing to a more meaningful study of patients getting protection from a complete nutritional support program and chelation therapy (where controls are kept just on the AAMP nutritional support program). The latter study, we believe, will show an 80 to 90 percent reduction in overall heart attacks in any group, versus the relatively small 25 percent reduction seen with Inderal in highly selected cases. The big difference is that the chelated patients will be fully active and feel great, while the medicated group on Inderal will frequently suffer significant side effects.

The arterial dilating agents cannot stretch, or make larger, a blood vessel that has a calcium-lined wall. The arterial wall is hardened, diseased, and scarred. Dilators will work only on those arterial segments that are undiseased. Their action, therefore, is mainly psychological. They have a placebo effect for the patient and give the physician a little easing of his own conscience—at least he tried to accomplish something—anything—for an otherwise hopeless disease. Most physicians finally realize, sadly, that they are unable to do very much for their arteriosclerotic patients. This realization is very hard on the average doctor's ego, as it reminds him of how truly impotent he really is. Eventually your physician must find the courage to join the movement toward chelation medicine.

Those physicians who are conscientious and empathic often have the compulsion to get the hopeless patients with really severe angina off their hands. Also, many doctors don't know the true statistics of failure for bypass surgery. They treat severe heart disease cases with standard drugs whose potential benefits are extremely limited. Such frustrated physicians are finally forced to sweep these personally discomforting patients under the therapeutic rug—out of their practices and into the hands of vascular surgeons or chest surgeons or others who perform open heart surgery.

These guilt-ridden physicians are also afraid of potential malpractice suits. The doctor may fear making a decision against bypass surgery for his patient, whom he knows is exceedingly ill and has a high risk of sudden death. The physician could later have the family dissatisfied, or even sue, on the ground that he did not do everything that could have been done. Thus, often reluctantly, a family doctor orders the "consultation" with the bypass surgeon, who has, of course, a strong financial motivation to sell the patient on the surgery.

Frequently, it is angina that doesn't respond to drugs that forces your doctor into ordering the arteriogram. If these X-rays show any significant blockage, most doctors feel they have to let a heart surgeon see you, just to protect themselves.

Another major reason for heavy referrals to open heart surgeons is the better financial return for referring cardiologists. They collect larger and more frequent fees to carry out preoperative and postoperative procedures on surgical in-hospital patients than to manage strictly medical therapy outside of the hospitals.

If physicians were more selective about their referrals and sent only the sickest patients for surgical repair, such as those very few whom chelation or calcium channel blockers did not help, the bypass procedure would not be so controversial. But the heart surgeons are watching their death rates and usually are rather careful to take only the more healthy candidates, such as forty-year-olds with a little chest pain and a slightly irregular electrocardiogram. They generally prefer not to operate on a desperate patient who is very likely to die from the surgery and thus increase an individual surgeon's death score, which the hospital doesn't like to see get too high. In contradiction to everything good and *pure about medicine*, that kind of seriously ill patient is denied surgery and seldom is informed of the chelation alternative. The only answer is a shrug of the shoulders by the surgeon and his admonition, "You are not a surgical bypass candidate." The really desperate patient is left to wait and wonder and put his family affairs in order. Sometimes he's even asked to donate his body to medical science. Even someone previously accepted for heart surgery is often refused for later repeated bypass attempts, after his graft reobstructs, no matter how seriously ill he is now. The heart surgeons guard their patient mortality rates against expansion.

Open heart surgery, then, must be considered the medical "garbage heap" for people with repeat problems. The surgical approach is usually abandoned by either the patient, who perceives it to be useless to go through it again, or the surgeon, to maintain his more favorable statistics. The surgeon usually abandons repeat procedures after no more than two operations, although one patient had the surgery *six* times. The patient usually simply dies, or lives as long as he can with his limitations[25,26]—unless he is lucky and hears about chelation, which often can still help him significantly.

THE CHELATION THERAPY CONTRAST TO OPEN HEART SURGERY

Hardening of the arteries is not a segmental disease. Medical science has shown in repeated studies that the condition does not affect only individual arteries, such as the coronaries around the heart, the carotids leading to the brain, or the aorta and femorals leading to the legs. It affects the entire body everywhere, particularly wherever an artery sends an offshoot to another part of the body.

Unlike surgery, which cannot be done on the smaller and frequently inaccessible blood vessels, the arterioles and the capillaries, chelation therapy does reach the tiny clogged segments along with the larger occluded ones. The entire arterial system and all the cells of the body have toxic minerals removed by EDTA infusion. Basic causes are thus treated this way. The load on the heart is reduced by lowering the resistance to blood flow. Anytime flow resistance in the arterial system is reduced, the heart is required to work less hard to accomplish the same job. Other problem areas improve, including a lowering of insulin requirements for diabetics, a drop in elevated blood pressure, a fall in excess serum cholesterol and trigylcerides, reduced senility, improved immune system functioning, more mobility for arthritic joints, improved kidney and liver function, and many other improvements.

Bypass surgery performed in the limbs, the heart, or anywhere in the body comes nowhere near providing the beneficial effects of EDTA chelation therapy. Chelation is known to prolong life and does heighten the quality of life. Its cost is usually less than 10 percent of the expense of bypass surgery.

Chelation should be the emergency treatment of choice when anyone has a heart attack and is in the intensive care ward being monitored around the clock by nurses. But that's not the way it is as yet. This life-saving treatment continues to be overlooked or ignored by the medical establishment, to the great disadvantage of the patient. It is now known that heart cells take many hours to actually die, and that lowering the intracellular calcium levels significantly decreases the number of cells that will actually die.

Following the symptoms of a heart attack, abnormal oxidative phosphorylation is occurring in the affected heart muscle, which is starving for nutrition. The energy that keeps your heart cells working falls markedly, similar to how the fire in your automobile cylinders won't move the car's pistons if the size and force of their explosions are too small. With a developing heart attack you will have insufficient energy even to get out of bed to go to the bathroom. Such a circumstance can be rapidly changed back to normal with chelation therapy.

In the mitochondria of the heart cells, the glucose and oxygen that you take into your body by eating and breathing are converted to adenosine triphosphate (ATP) from adenosine diphosphate (ADP). ATP supplies your heart and other body muscles with energy. Without it, you won't have much of an explosion in your own metabolic cylinders to keep your body engine running. It will be as if you're turning the crank by hand, looking for strength without any automatic choke on the accelerator pedal, and you will have run out of gas.

As the cells become starved for energy, their pH changes from neutral to acid. An altered pH traps calcium ions and draws them to the cells to further block oxidative phosphorylation. (Oxidative phosphorylation is the formation of "high energy" released by the burning of various substances in the body.) This continues to prevent the conversion of energy. It's a vicious cycle of chemical circumstances feeding on themselves. But a remedy does exist. All that your cardiologist has to do is give a calcium chelating agent, and the heart attack symptoms will usually be corrected almost instantly. The mitochondria will function again; ATP will rise back to normal; cell metabolism will return to where it was before the heart symptoms began. This has been proven and published by four medical investigators at the University of Arkansas School of Medicine.[27]

The four researchers—two cardiologists and two physiologists—have shown that some form of calcium chelating agent (it doesn't have to be EDTA) will restore the heart. If you have sustained an acute myocardial infarction, you have not yet killed a piece of the heart muscle. A true heart attack has not actually struck. Instead, the heart is having total starvation of a particular section and is sending a sharp message. "Hey," it's warning, "do something quick to get me oxygen and energy!"

Any one of the chelating agents, such as ascorbic acid, Ringer's lactate (containing lactic acid), or other weak organic acids acceptable for intravenous infusion, will help reverse the process of heart muscle disintegration and protect the cells against dying, clots from forming, and arrhythmias from developing. Of course, EDTA works most swiftly and effectively; however, a cardiologist unfamiliar with its application can employ another safe chelating agent.

Calcium channel blockers aren't as efficient in permanently restoring heart health as is EDTA chelation therapy, but even these calcium antagonists are clearly better as a coronary medical program than open heart surgery. They inhibit the excessive accumulation of calcium in the heart cells and allow ATP production. Additionally, if you are the patient in heart spasm, you can help avoid death of the starved portion of your heart muscle. You won't show the elevated enzymes (CPK, LDH, SCOT, and others) that your doctor measures in your blood test each day to see how many heart cells really died and released their enzymes. An actual heart attack will be avoided; only a severe attack of angina will have occurred. You will usually be able to go home from the hospital the next day by having calcium blocking agents and/or chelation therapy administered to you. Even as you lie in a state of anxiety, let's hope you can keep control of your own situation.

Sometimes keeping control of what's done to you is impossible. If the hospital's electrocardiogram indicates you are starting to have an acute heart attack, your need is to have a bottle of EDTA or even lactated Ringer's with 10,000 mg vitamin C and 1000 mg magnesium chloride added for intravenous feeding plugged into your vein. With the health professionals taking another EKG in four hours, usually no evidence of myocardial infarction will be seen. Frequently the EKG now appears entirely normal. An immediate follow-up with more chelation therapy and magnesium and the complete nutritional support program must be instituted on the following days, but you usually don't have to stay in intensive care or occupy a hospital bed, except possibly for one night just for observation and rest.

Nowadays, what's the usual procedure when you enter the hospital emergency room with an apparent heart attack? The health professional personnel place you in the cardiac intensive care ward and drip fluids into your arm, even when such fluids are technically unnecessary, in order to keep the lines to blood vessels open. They are anticipating that an acute arrhythmia or rhythm disturbance, where your heartbeat starts ineffectually racing, will affect your heart. A heart muscle beating 200 times a minute is equivalent to a zero heart rate. To offset such an arrhythmia, the personnel must shoot in appropriate drugs to keep up your blood pressure and usually will give electroshock therapy to convert you back to a normal rhythm. According to the information presented in this chapter, such an obsolete procedure is largely unnecessary. The calcium blocking agents prevent most arrhythmias, says the Chief of Cardiology at the Lahey clinic in Boston.[28]

In my opinion, chelation therapy as an emergency measure included in the cardiac medical program should practically empty hospitals of their occupied beds in the cardiovascular wards. Health care costs will come down and lives will be saved. Even Social Security's Medicare program could save enough money by not having such heavy occupancy of hospital facilities. Indeed, use of the treatment might help support the Social Security system.

You, as the patient having heart attack symptoms, should demand that citric acid, vitamin C, Ringer's lactate, and/or EDTA be dripped intravenously into your vascular system. Tell the emergency physician, "Hook me up to a chelating agent and measure my ionic calcium as you're dripping it in. You will see the ionizable serum calcium drop to half the normal. Inject me with the chelating agent, Doctor, because if not, I'm going to sue you and this hospital for medical malpractice!" You'll help to turn around the practice of emergency cardiac medicine in the United States and head it in the right direction. At the same time, you'll be saving your own life.

CHAPTER TEN

Health Insurance Industry Opposition to Chelation Therapy

Signs that his arteries were hardening first appeared when Gene Zaro of Stamford, Connecticut, was seventeen years old. A lump the size of a marble appeared in his left leg. The lump was red and painful and lay just beneath the skin surface of the calf. The boy consulted Barnet Fine, M.D., of Stamford, who was then his family physician. Dr. Fine could not determine the origin of this nodule, but it soon disappeared.

Lumps came again in his leg when the young man was twenty years old. Gene was hospitalized with a diagnosis of phlebitis and scheduled for an operation. Two surgical incisions were made to remove the new nodules, but the surgeon did not see the presence of phlebitis and simply closed the wounds instead, with nothing accomplished. Again, the signs of disease went away by themselves.

Five years later small subcutaneous lumps appeared on the back of the man's left hand at the base of two fingers. Staff physicians of Stamford Hospital were completely baffled by the abnormality. For three months Gene Zaro suffered with these bluish-red, deep-seated bumps that tingled and felt tender and painful. They interfered with his work as a house painter, but he tried to ignore them. Finally, they faded away.

In June 1972, severe pain hit Zaro's left leg, along with a recurrence of larger red lumps. His foot and leg felt cold, numb, and tingling. He consulted Arthur Kotch, M.D., at Danbury Hospital in Danbury, Connecticut, who diagnosed his condition as nodular vasculitis, a peripheral vascular disorder characterized by recurrent nodular nontuberculous lesions beneath the skin, especially in the legs. The nodules are sites of obliterative inflammation of the arteries and veins.

At the Danbury Hospital, corticosteroid injection treatment with prednisone produced several bad side effects for the patient. He experienced high blood pressure and a subsequent breakdown of the nodules into irregularly shaped ulcers that gave off a watery discharge. However, biopsies of the lesions revealed nothing in particular to the pathologist.

Zaro next looked for relief in the Waterbury, Connecticut, Hospital. Waterbury Hospital was closer to his home in Brookfield, Connecticut, where he had relocated for a time with his wife, Sue, and their four small children. Redness, tingling, burning, and feelings of deep pain from the lumps in his left calf stayed with him constantly. They were disabling and kept him from work. He broke out in a skin rash also. No change in diagnosis or treatment was offered; consequently, the patient tried the Bridgeport Hospital in Bridgeport, Connecticut. Nothing new or relief-giving was forthcoming from that source either.

For two years Gene Zaro stoically suffered with symptoms and signs of his peripheral vascular disease. Let me differentiate between symptoms and signs. The term *symptom* describes any morbid departure from the normal in function, appearance, or sensation, experienced by the patient and indicative of disease. A sign is any abnormality discovered by the patient or by the physician at his examination of the patient. Thus, a symptom is subjective and a sign is objective.

When Gene Zaro could bear the symptoms of searing pain no longer, since each lump felt like it was being scorched with a hot iron, he entered Yale–New Haven Hospital in New Haven, Connecticut. There, the physicians started from scratch and disregarded whatever had been diagnosed or done before. However, nodular vasculitis was reconfirmed as the condition, for which cortisone therapy was tried again. As before, it did not work.

Another year passed while Zaro moved back to Stamford and underwent more biopsies and his first angiogram. It showed sporadic blockage from structural changes in the walls of the arteries of his left leg. The intima was roughened, and blood clots attached to the vessel wall with obstruction of the lumen had appeared. Zaro's arterial lumen was markedly narrowed and almost entirely clogged. He was suffering from the spasms of calcification in the walls of his arteries.

The physicians at the Yale–New Haven Hospital said there was no *medical* relief known for this patient's condition. They mentioned nothing about chelation therapy. They talked, instead, about surgical bypass of the femoropopliteal arteries, but did not recommend the procedure at that time. A femoropopliteal bypass is an operation in which a vascular prothesis is used to bypass an obstruction in the leg's femoral artery. The graft may be synthetic material such as Dacron, or it can be the patient's own saphenous vein, or it may be the carotid artery taken from a cow.

Symptoms of stabbing pain and signs of recurrent nodules were chronic for this victim of arterial occlusive disease. He limped with intermittent claudication of both legs, with pain much worse in the left. The lumps came up, burned and tingled, and occasionally went away. Some lumps stayed, while others that had disappeared might reappear in the same spots. All of them gave Zaro agonizing discomfort. Pain kept him awake, and sleep came in fits and starts.

In the summer of 1975 the patient, now thirty years old, banged the big toe on his left foot. The toenail turned green and the toe tip seemed opaque. The skin at the site of injury became darker and finally blackened with gangrene. The Yale-New Haven doctors performed another angiogram on Zaro in November 1975 and recommended immediate bypass surgery. He consented to subject himself to surgery. On the operating table, the surgeons saw the deteriorated condition of his arteries with their excessive blockage. They realized the bypass procedure was useless and that an amputation was needed instead.

The doctors finally concluded that the man really had Buerger's disease, thromboangiitis obliterans, which is a circulatory disease of unknown cause. It is recognized, however, that excessive smoking is an important factor. (The patient had been smoking a pack and a half of cigarettes a day and ate a diet high in refined carbohydrates such as sugar.) Injury often precipitates an acute episode of Buerger's disease, with rapid development of gangrene.

Gene Zaro had shown many of the signs and symptons of Buerger's disease, starting from age seventeen. Typically, the problem occurs in young males, with an onset that is gradual. The signs and symptoms come and go in an irregular pattern until they cause gangrene and eventual death. The man had complained of coldness, numbness, tingling, or burning. There was a history of migratory phlebitis and intermittent claudication. He experienced very severe pain or tightening after muscular exertion and redness of the foot and leg.

The Yale-New Haven surgeons did not amputate Zaro's left leg right away. Rather, they gave him the opportunity to make his own thoughtful decision. They were quite sure he would elect to have it done. The patient left the hospital for the Thanksgiving holidays and remained with his family for two days. His leg pain was so unbearable, that he returned to the hospital. They first amputated at the ankle but then moved to within three inches below the knee because arterial blockage was apparent farther up in the calf.

The former house painter recuperated and was fitted with an artificial leg. His doctor kept close check on the stump's healing by means of Zaro's monthly or bi-weekly visits.

When, on February 14, 1977, Zaro complained of low-grade pain in his right leg, his physician ordered a third angiogram. In shock, the patient and his wife learned that his right limb was in danger also. The doctor advised that a surgical bypass was needed, but the distressed couple decided to wait and think over very carefully whether to undertake this second operation.

The nodules began to appear in his right leg, and it looked to Mr. and Mrs. Zaro as though the bypass operation was inevitable. But the surgeon changed his mind when he reexamined the patient, and instead recommended a sympathectomy. A sympathectomy is the excision of a local segment of the autonomic nervous system that prevents the damaged arteries from demanding more blood for the leg. With the nerve supply blocked, the pain would not be produced. Nevertheless, the effect is only temporary and finally must lead to leg amputation because of the potential onset of gangrene. The patient refused to have this procedure.

Gene Zaro just waited for what would happen next. He knew he had Buerger's disease with its accompanying poor prognosis. The focus of his worry then switched from the remaining leg to his two arms where egg-sized lumps had popped out. His arms felt as if they were sunburned raw in spots where nodules bulged under the skin. Much of the time he had to sit with only his hands resting on the arms of a chair. He was able to do little more than watch television all day long—and wait—and hope.

It was at this point that the Zaro family heard about the chelation answer to Buerger's disease and other manifestations of hardening of the arteries. Gene Zaro made an appeal for chelation therapy to be administered to him at the nearby Veterans Administration Hospital. He addressed his verbal and written requests to the regional VA office that had been paying the medical bills for much of his previous care—a sum amounting to many thousands of dollars. Upon checking with the physicians and surgeons at the VA hospital, the regional VA office responded that chelation treatment for Gene Zaro was denied.

The third-party health care provider readily agreed to pay to have Zaro's remaining leg cut off, or to pay the costs of his surgical sympathectomy (whose benefits are uncertain at best for his condition) but declined to cover a medical injection procedure for saving his limb and possibly his life. The same problem with insurance payment or reimbursement is generally encountered by most patients with vascular disease when they want chelation treatment rather than some other, frequently unproven, life-endangering surgical procedure that is being recommended to them by their local specialists.

A peripheral vascular surgeon can advise his patient who has partial gangrene of a foot that he must undergo amputation of the leg below the knee. The surgeon will receive no criticism for such a proposal from his colleagues, the hospital staff, the patient's health insurance carrier, the doctor's medical society, his state medical board of examiners, or anyone else. But let a chelating physician suggest that the patient not have his leg removed and undertake chelation therapy instead to save it, and the chelator opens himself to vilification by almost every person or institution involved with rendering or reimbursing for traditional medical care.

Chelation doctors can't help everyone, but in 200 patients taking chelation therapy for reversing gross gangrene, cases similar to that of Roland Hohnbaum (described in Chapter 2), only 3 people lost a limb. Obviously, all of the gangrenous legs in this study from the AAMP files didn't get well, but saving 197 limbs is a lot better than cutting off all 200.

Why wouldn't all the people in danger of amputation because of gangrene take chelation therapy rather than have their limbs removed? There are two reasons: (1) Most people seem unaware of this chemical treatment's existence, and (2) third-party health insurance carriers are producing a serious disruption in the marketplace for medical care—patients often fail to get sane, noninvasive, and safely administered chelation treatment in the doctor's office because third-party carriers deny reimbursement. They limit the therapeutic competition among doctors this way by only reimbursing for certain therapies such as surgery. A clear-cut restraint of trade among health professionals, which I shall further clarify in the next chapter, forces health insurance policyholders, restricted by their financial circumstances, to choose among the few approved forms of therapy that get paid. This happens because the doctors who control organized medicine tell the insurance companies what they classify as the "usual, customary and reasonable" treatments.

The physician-politicians of organized medicine who have intimate relationships with regional Blue Cross–Blue Shield insurance companies see to it that health insurance guidelines don't allow reimbursement for therapies that are threatening to their own income. What physician-controlled Blue Cross–Blue Shield sets as payment policy is usually followed closely by all the other private third-party health insurance carriers. By disallowing payment for chelation therapy but covering nearly 100 percent of the cost of competing therapies, including amputation, health insurance companies force most financially pinched patients into having less than optimal treatment. All too frequently, their diseased legs get cut off, when chelation could have avoided needless pain and suffering and usually saved the insurance company money at the same time. All of us can only ask, *why*?

THE "COHESIVE CONTRACT" HEALTH INSURANCE GAME

The money going into the insurance pool (including your payment of Social Security premiums) is collected from everybody. The consumers making payments are told that they are entitled to Medicare or other health insurance. However, these premium payers are hardly ever advised that coverage is provided for essentially standard AMA-approved treatment and certain types of innovative therapies are not included in coverage. Sure, the state insurance law requires that the policy must lay out its limitations, including the usual noncompensation for pregnancy bills or psychiatric care. These were standard exclusions for years.

What you, the health insurance policy purchaser, fail to realize or choose to ignore is that you have to accept the policy as given. You can't add two extra words for coverage clarification. But you're never informed that you are buying an adhesion-type of contract, which means you are not free to alter the conditions the insurance company imposes. Thus, without specifically being informed of it, you have entered into an agreement that precludes your visiting health practitioners outside of the medical mainstream to receive new or innovative, alternative, and competing forms of therapy. You don't really have any freedom of choice anymore as to what kind of doctor will deliver your medical care or what that medical care will be. You must take only "customary and usual" or "reasonable and necessary" care, or pay for it yourself. The health insurers get to decide what constitutes usual, reasonable, necessary, and customary.

The dispensing of medical services is one of the largest businesses in the United States. In calendar year 1981, the American people spent just short of $250 billion on health care, about 9.4 percent of the Gross National Product. This huge figure represents over a 100 percent relative increase since 1950, when the total outlay amounted to about $13 billion, or 4.5 percent of the GNP. Currently, government covers over 40 percent of total health outlays, and insurance—commercial carriers and the Blue Cross-Blue Shield plans together—almost 30 percent. The consumer pays out of pocket only about 30 percent, and that goes mostly for drugs, nursing home care, dental care, and payments to physicians for routine office visits. When it comes to the most expensive item, hospital care, currently amounting to about $100 billion annually, the consumer pays out of pocket only about seven cents on the dollar.

Tucked within all this payout is the cohesive contract of the health insurance industry, which represents a trick played on the medical consumer. Health insurance companies play a nefarious game with their health insurance cohesive contract, which was designed to control costs for the health insurance industry. Unfortunately, it surpresses competition in the marketplace. It prevents compensation for medical services from filtering into alternative methods of healing. Under the guise of avoiding medical quackery, the cohesive contract frequently holds back the dispensing of legitimate and effective physician services. By allowing payment only for establishment medicine, the cohesive contract causes significant minority viewpoints among doctors to be strangulated. If you have wondered why it often takes thirty to fifty years for a healing technique to be adopted by American medicine, this is one of the major reasons. Minority medical groups have their platforms cut from under them by the health insurance industry, which lists alternative therapies not popular with the majority medical groups as nonreimbursable.

Chelation therapy falls under the exclusion listings in a cohesive contract. The health insurance industry is committed, as much as possible, to preventing policyholders from being reimbursed for chelation treatment.

My personal opinion, based on extensive interviews of physicians and patients, is that health insurance companies seem to fear the unknown. Even though it appears certain that they would save much money over the long run with new developments such as chelation therapy, they worry about inordinately large numbers of people initially taking chelation treatments. Their fear is real inasmuch as until now little or nothing could be done to prevent or reverse hardening of the arteries. With people suddenly becoming aware of a rejuvenation program that works, the conjecture among preventive medicine specialists is that insurance companies fear the excessive start-up costs. EDTA injections paid by health insurance could soon save millions of dollars as the hospitals began to be less crowded.

Chelation therapy is the ultimate preventive medicine against degenerative disease. Health insurance companies today simply aren't encouraging preventive medicine because they aren't convinced it will be cost-effective for them. Your health insurance won't reimburse you for a medical checkup. They're willing to reimburse you only after you've developed serious disease. Insurance pays you for lost health when you're attempting to salvage whatever you have left. Get really sick—become obviously afflicted with advanced blood vessel disease; fight angina, gangrene, or senility—and you may collect something. The companies do not encourage you to have advanced noninvasive, painless vascular tests to identify the disease in its early stages, and then take adequate treatment to reverse or at least control it.

However, if a prospective insured would agree to undergo complete noninvasive studies, the insurance companies could much more actively reward healthy patients by offering premium reductions. In essence, this would be paying patients to stay well. Such a concept has been tested in California with health insurance in which employees who did not get ill, and thus did not have significant medical expenses, received up to $500 a year in rebate from their health insurance. Many more people began jogging, lost weight, and stopped smoking and eating sweets than anyone could have dared to dream—and illness rates dropped dramatically. In essence, everybody benefited! I hope these proposals are soon tried elsewhere.

The philosophy of health insurance carriers has always been that they should sell policy coverage only for something that happens to just a few of us each year. They must be panicked by a new procedure everyone over forty-five might want performed for the reversal and/or control of their potentially serious hardening of the arteries revealed by noninvasive tests. As the insurance companies see it, holistic doctors tell me, paying for chelation therapy would cause a negative cash flow for the companies over several years. They refuse to recognize how quickly they would see a major drop in heart attack, stroke, and even cancer rates. These drops could provide both short- and long-term financial savings from just reducing strokes and heart attacks among their policyholders by as much as 50 percent or by reducing the incidence of cancer by as much as 60 percent.

The dramatic decreases in various other degenerative diseases that have been reported could offset the start-up chelation expenses very fast. Insurance executives appear excessively concerned regarding the possible short-term excess expenditures from the reimbursement of policyholders taking preventive chelation injections, even though the evidence shows that most of these patients will avoid subsequent hospital care for years.

Most reasonable people will want to take advantage of a major breakthrough in the treatment of vascular disease, since 90 percent of us eventually develop this disease. It eventually contributes to the death of 54.7 percent of the population of the United States. The insurance industry has not yet developed the mechanisms for dealing with the negative cash flow that may initially develop if this treatment became widely employed.

Additional savings would be found from reducing by at least 75 percent the number of open heart surgeries done today, which cost $20,000 to $30,000 apiece. In contrast, for every coronary artery bypass operation performed today, approximately a dozen heart patients are each able to receive a series of twenty-five to thirty chelation treatments. The costs are nearly equal: one surgery patient to a dozen chelation patients.

As a health insurance policyholder, therefore, you should realize that your purchase of health insurance from all private health insurance companies leaves you with major unstated restrictions. The same is true for all of you who have money withheld by the Social Security Administration for Medicare. You are being forced without actually being informed about it into accepting standard establishment-type medicine, which may vary considerably from what you would have chosen or from what your doctor actually might want you to receive.

We now have Blue Shield physicians who agree never to bill the patient but to accept as full payment whatever Blue Shield pays them for your health problem. These physicians can never try an innovative therapy, because it won't be paid by Blue Shield. Blue Shield compensates doctors only for usual, reasonable, necessary and customary care and labels everything else experimental or unproven.

HEALTH INSURANCE COMPUTERS JEOPARDIZE MEDICAL LICENSES

Your insurance claim, filed by you in your local community, may jeopardize your physician's license to practice medicine. Anytime your doctor gives you innovative treatment such as chelation therapy he puts his license on the line. This occurs whenever he varies from standard practice, which I will explain. Then the physician will be looked upon by his Professional Standards Review Organization (PSRO) as incompetent or engaging in unprofessional conduct. This problem is only now becoming a nationwide threat. It has taken years for the insurance industry to put into operation all the procedures and controls on medicine that may soon make it possible to entirely control its practice. Most doctors are completely unaware of the dangers that I am alerting you to in this book. Chelation doctors were among the first to be singled out for attack, and all other nonconformist physicians will soon become aware of what I am saying here.

The effect is that your own freedom of choice in therapies—the right to make the final decision on how your health problems should be approached—is rapidly disappearing. Not only do you fail to get reimbursed for medical care for which you are paying out premiums, but now your physician is liable to lose his medical standing, his medical license, and even his livelihood simply because he tried a modality that is not yet in the medical mainstream. Innovative treatment is beginning to be stifled and soon may be completely eliminated in the practice of American medicine.

How does such an action take place? The first thing to happen is that the doctor's insurance claims from all of his patients are put on "full peer review." This means that even his usual and standard treatment is questioned in every request for health insurance reimbursement.

As the patient, your claim for even ordinary and well-accepted procedures such as an X-ray for a minor injury are peer reviewed and held back from being paid. The delay is sometimes for months and sometimes forever because the carrier may arbitrarily *deny* payment entirely. It doesn't matter that your doctor is a Blue Cross–Blue Shield participating physician and has agreed to accept the Blue Shield payment schedule as full payment; you still wait for final disposition of the bill. Whether you've been relieved of an ingrown toenail, a headache, or something major, Blue Shield's payment is the participating physician's entire fee. More than one innovative physician has been nearly bankrupted when insurance companies withheld all of their patients' reimbursement checks so that the patients were unable to pay their bills in a timely fashion.

Now, if he doesn't float along with the medical mainstream, how is the doctor going to receive steady payment of his Blue Shield and other insurance fees? Inasmuch as any health professional using innovative methods is sure to be identified by the health insurance company computer, the doctor learns to take no chances. He attempts to maintain his practice in a noncontroversial way. To assure that his claims are not thrown into the "full review" pile for peer review, the doctor not wanting to be hassled makes sure that he utilizes only standard and traditional middle-of-the-road medical care acceptable to the health insurance company. Even if he might want to try vitamins, chelation, or some other medical alternative for your family's health care, he's not likely to do it.

This entire system of health insurance reimbursement tends to remove from the average physician any incentive to give you the very best medicine. Taking the risk of recommending or giving treatment that's different can get him identified by the insurance claim computer. His income could be drastically cut and his medical license endangered. Your medical champion—the person you rely on to try anything safe and effective to make you well—disappears when your doctor accepts full Blue Shield, Medicare, or other health insurance payment.

There are doctors today who are the ultimate examples of closet chelators. To avoid full or partial peer review, some AAMP physicians accept patients for chelation therapy only on the condition that they will not file health insurance claims for possible reimbursement of their medical bills. This further restricts dispensing chelation services, since many patients do refuse to finance their own treatment when they are already investing large sums in annual health insurance premiums. This is medical restraint of trade of the highest order. It cuts the patient population to just those relative few who can afford to pay for the chelation treatment out of their own pockets.

How medicine is being practiced in America is largely determined by the payment schedules formulated in potentially collusive meetings among health insurance companies. The Prudential Life Insurance Company throughout the 1970s paid all claims submitted by anyone who underwent chelation therapy. Then a patient in Sacramento, California, who was covered under the Blue Shield of California insurance plan filed for chelation care reimbursement. He used the example of Prudential's compensation coverage of another patient with a similar vascular problem as a good reason for his own medical bills to be paid by Blue Shield of California. The Blue Shield plans across the country are largely physician-controlled, as I've pointed out, which was the case with Blue Shield of California. Consequently, after an extensive exchange of correspondence and meetings between representatives of the two companies, Prudential Life Insurance Company letters went out to all reimbursed health insurance policyholders saying that chelation therapy bills had been paid in error and the company wanted its money refunded.

It's obvious to anyone studying these actions that organized medicine has too much control over the disbursement of health insurance funds. Physician politicians and health insurance industry executives are in bed together, illegally restraining free trade in the practice of medicine in the United States.

Administrative Law Judge Thomas F. Howder ruled on a case for the State of Michigan on June 30, 1981. He ruled that the Michigan State Medical Society conspired with some of the physician politicians and constituent medical societies to fix fees and regulate reimbursement policies of third-party payers. His decision upheld a Federal Trade Commission administrative complaint, filed in August 1979, that the medical society violated federal law by restraining competition among physicians. The judge rejected the society's argument that it is exempt from antitrust scrutiny.

In his ruling, the judge said that the Michigan State Medical Society had tampered with physicians' fees, engaged in "concerted action" to regulate cost containment and reimbursement of third-party payers, and engaged in illegal "concerted negotiations" concerning these policies.

The FTC's 1979 complaint alleged that medical consumers had been deprived of the benefits of third-party independently determined reimbursement or cost-containment efforts.

In his opinion, Judge Howder said that "anticompetitive effects and consumer injury did result from [the medical society's] actions. Michigan

Medicaid increased its reimbursement to physicians. Differential reimbursement to doctors in different regional areas was eliminated at a cost of 'millions of dollars' to [Blue Shield]. As a matter of logic, this will lead to higher [Blue Shield] premiums [to consumers]."

To ensure that there would be no mistake about interpreting the ruling among doctors and that the restraint of trade in Michigan medicine would be discontinued, the Michigan State Medical Society was required to make an extensive communication effort. This medical society had to mail a copy of Judge Howder's eighty-three-page decision to each of its component societies, each of its specialty sections, and each of its 8700 members within thirty days after the order was final. The society had to also publish the judge's ten-page order on the first pages of an issue of *Michigan's Medicine*, the state medical journal, within sixty days and on the first page of the next issue of its bulletin *Medigram*. For the next ten years, the medical society must provide each new medical society member with a copy of the decision when the new physician is accepted into membership.[1] Of course, the Michigan State Medical Association is appealing this decision.

ORGANIZED MEDICINE USES MEDICARE AGAINST COMPETITION

The U.S. Supreme Court has heard the matter of the *American Medical Association* v. *The Federal Trade Commission*, one of only 23 cases selected from over 2000 applying for a hearing. The Supreme Court has also recently decided a case entitled *National Gerimedical Association* v. *Blue Shield.* Both of these cases go to the issue of the applicability of concepts of antitrust law to the field of medical care. In *National Gerimedical*, the court decided that medical care is subject to antitrust regulation. In the *FTC* case, on AMA physician advertising, which the court ruled on in the 1982 session, the justices dealt with the issue of the applicability of federal governmental regulation to the same area.

The local Blue Cross plan in Kansas City, Missouri refused to sign a contract with the National Gerimedical Hospital and Gerontology Center of Kansas City, claiming that the facility was built contrary to local health planning standards set up by the municipal planning agency. The hospital sued, contending that the absence of a Blue Cross contract would make it more difficult to compete for patients.

In an earlier decision, a federal judge, who was later upheld by an appeals court, had said antitrust laws didn't apply to the relationships between Blue Cross and the planning agency. The lower courts said the insurer and planning agency were acting to carry out federal policy as spelled out in the national Health Planning and Resources Development Act.

A persistent legal issue in health planning has been whether actions by organizations such as Blue Cross, in concert with health planning groups, violate the antitrust statutes. The U.S. Supreme Court has now ruled that they do. The Justice Department had sided with the National Gerimedical Hospital and Gerontology Center, claiming the Court of Appeals' decision "had a substantial adverse impact on competition and antitrust enforcement in a large sector of our economy." The Health and Human Service Department, the administrator of Medicare, had sided with Blue Cross and Mid-American Health Systems Agency, the planning unit involved in the lawsuit. Now that the Department of Health and Human Services has backed the losing side, it is going to be closely scrutinized, and a key case initiated by Leo J. Bolles, M.D., is in the courts awaiting decision right now.

The Health Care Financing Administration (HCFA) U.S. Department of Health and Human Services sent Leo J. Bolles, M.D., of Bellevue, Washington, a letter on July 25, 1978, informing him that he is going to be excluded from the payment program for Medicare recipients. The excuse given was that Dr. Bolles was found to have "furnished services that were harmful and substantially in excess of individual needs." During the course of the physician's subsequent 1981 trial, which he had requested by bringing suit against the HCFA, it came out that this is the way the health insurance company's computer describes EDTA chelation therapy.

It was also claimed by the HCFA that this chelating physician was doing too much of "all other procedures," meaning he was performing noninvasive vascular testing before and after the patient's treatment with chelation. Dr. Bolles was correctly following the protocol of the AAMP to assure safety and provide progress reports for the chelated patient. The computer identified the use of this patient testing procedure as excessive because it compared his use of these tests and treatments to general practitioners and other physicians who seldom, if ever, order noninvasive tests. Non-holistic health professionals don't know much about noninvasive tests. Since they don't do chelation, traditionalists would have little use for the information learned anyway. In Bolles's case, the HCFA apparently wasn't even cognizant of the way chelation therapy is given—by intravenous injection.

The actual exclusion of Medicare coverage for the patients of Dr. Bolles was levied on October 10, 1979. According to Dr. Bolles, organized medicine used Medicare as its instrument of vendetta. Dr. Bolles had resigned in 1973 from the King County (Seattle area) Medical Society. He had serious and bitter clashes with this organized medical group going back to 1971. Then, as now, he had been criticized and attacked for utilizing metabolic and nutritional approaches to degenerative disease and chelation therapy for hardening of the arteries. "I will not cease and desist in what I am doing," Dr. Bolles declared. "I don't feel I'm doing anything ethically or morally wrong. I'm practicing according to preventive medicine standards." The Seattle Medical Society did not accept Bolles's declarations in 1971, 1973, or now. Then, physician colleagues finally moved against him when they discovered that the "profiles" of medical practice patterns derived from Medicare claims submitted by patients and analyzed by Medicare's computer could be their instrument to stop Dr. Bolles and his colleagues from using chelation therapy.

In the years 1976 to 1981, Dr. Bolles had provided chelation therapy for 4000 new patients. "For the most part, they are doing quite well," testified William Correll, Business Administrator of the Bolles clinic. "I would estimate that less than 1 percent of his patients have not been fully satisfied with their cardiovascular improvement."

Twenty-two patient history records were reviewed in the Bolles clinic on behalf of Medicare by a Washington State University vascular surgeon, D. Eugene Strandness, M.D. Dr. Strandness testified as a prosecution witness that after reviewing all the "before and after" noninvasive studies performed by Dr. Bolles, he could not find any evidence of hardening of the arteries in these people either before they received chelation therapy or afterward. In an interview, Dr. Bolles claimed that all of the patients Dr. Strandness examined were over the age of sixty-five and some were in their late eighties. One had recently undergone bypass surgery, and some had been referred to Dr. Bolles by other physicians because of their arteriosclerosis.

Dr. Strandness is coauthor of a textbook on noninvasive vascular testing, entitled *Hemodynamics for Surgeons.* It suggests that almost everyone by age fifty has arteriosclerosis and excess accumulations of calcium in their aorta. Yet Dr. Strandness's testimony indicated that Dr. Bolles had managed somehow to find twenty-two patients of advanced age who had no arteriosclerosis. Most of these patients had symptoms of impaired circulation and/or documentation of circulation on their noninvasive studies, which Dr. Strandness was asked to review. Dr. Strandness had contributed and helped edit his textbook that completely supports the use and validity of the tests Dr. Bolles was doing. Dr. Strandness did not wish to endorse at the hearing many of the statements contained in his own textbook, possibly because it tends to help support the validity of chelation therapy.

However, the blood vessel pathology was quite apparent in these twenty-two patients to five physicians viewing their pretreatment noninvasive test results. Among the experts at the trial testifying about the patients' marked blood vessel improvements were Elmer M. Cranton, M.D., President of the American Holistic Medical Association; H. Richard Casdorph, M.D.,

Ph.D., Diplomate of the American Board of Medicine, Fellow of the American College of Physicians, and Assistant Clinical Professor of Medicine, University of California, Irvine; Garry F. Gordon, M.D., then Chairman of the Board of the American Academy of Medical Preventics; and two other qualified physicians.

The decision on the Bolles case is likely to be handed down by U.S. Administrative Law Judge Keith Callow late in 1982. But it is only one of the latest incidents in which organized medicine appears to be involved in harassing alternative medicine practitioners, using Medicare's computers to force conformity in the treatment and diagnostic methods employed in American medicine.

MEDICARE BEGINS TO MAKE CHANGES

Correction of Medicare abuses of chelation therapy patients may come fairly soon if we can get an informed public behind chelation. A letter received January 2, 1981, by U.S. Senator Howard M. Metzenbaum, Democrat of Ohio, which was sent by Robert D. O'Connor, Director of the Bureau of Program Policy, the Health Care Financing Administration of the Department of Health and Human Services, states in part: " . . . effective May 15, 1980, Medicare's general policy on the coverage of drugs was changed so that any use prescribed by a physician of a drug approved for sale by the Food and Drug Administration (FDA), as is EDTA, may be covered if the Medicare contractor (Blue Cross-Blue Shield plans and other health insurance companies) that processes the claim determines it is '*reasonable and necessary*' for the individual patient and all other coverage requirements are met. This general policy is described in program guidelines issued to all contractors and will apply to chelation therapy until we receive a specific recommendation from our medical consultants and determine if a special policy must be applied to it. If so, Medicare contractors will be informed.

"Program guidelines provide for contractors to determine whether Medicare's coverage requirements are met in an individual case with the advice of medical consultants and to take into consideration the circumstances of the case and accepted standards of medical practice."

As a result of this Medicare rule change, some people are receiving reimbursement for chelation therapy if they underwent treatment between May 16, 1980, and October 1, 1981. After October 1, 1981, another rule took effect, which I shall describe later in this chapter. If you wish to receive reimbursement for expenditures made for chelation therapy and you were a Medicare recipient on May 16, 1980, or you wish to undertake chelation therapy in the near future and want to try for later reimbursement, send a facsimile of the letter published in Appendix III. Merely incorporate your appropriate personal information in the blank spaces provided.

Today chelation therapy may or may not be reimbursable by Medicare, depending on where you live and many other variables. It depends upon your local health insurance carrier, because the United States is divided into ten different districts for Medicare purposes.

Medicare Regulation 799 allows for local insurance companies that insure Medicare to elect to reimburse for the use of certain drugs that have previously been approved by the FDA in areas other than their labeled uses. This means that if a drug was approved by FDA for one particular use and your doctor believes this drug could be beneficially employed in other ways, then Medicare could reimburse for the use of the drug in other ways. However, this judgment is left up to each individual insurance company that handles the Medicare coverage. The decision must be approved by the company's staff of physicians—who almost invariably belong to organized medicine. In no event are insurance doctors willing to jeopardize their paycheck by making waves.

For intravenous injections of EDTA to be reimbursed to you as treatment of hardening of the arteries, a staff of physicians of the local Medicare insurance carrier must say that the use of EDTA in your case was "reasonable and necessary" and that its use meets the accepted "standards of medical practice."

This language was not explained to you when you purchased your insurance policy or paid money into Medicare. Yet these words will be used to deny you any reimbursement in most chelation cases. There are acceptable and useful legal definitions for "reasonable and necessary" in some states. Most of the time the legal definitions are ignored by insurance carriers, probably because the definition simply states that if your licensed health practitioner believes a treatment is appropriate for demonstrable disease, then it must be considered reasonable and necessary. Your insurance company would have to pay if it accepted this clear legal definition.

Across the country most chelation patients are routinely refused reimbursement by the insurance carriers and even by Medicare Special Administrative Hearing Officers on the nebulous basis that chelation therapy is not deemed by them to be reasonable and necessary. Such is the case even though every doctor may have previously told the patient that his condition was completely hopeless, and he is now well and even able to walk again.

The other loophole is the accepted "standard" of medical practice problem. One-third of all practicing physicians belong to the AMA. Although this is not even a simple majority, if the AMA doesn't endorse a treatment, it seems too often that the treatment cannot be deemed to meet the accepted "standard."

To ensure that all insurance companies will refuse new therapies on a more uniform basis across the country, Blue Shield and Blue Cross have developed their special "medical necessity" project list to help legally define what this nebulous acceptable "standard" is. It's done in such a way, of course, that chelation, hair analysis, megavitamin therapy, hyperbaric oxygen, and other innovations will not be able to meet their vague and ambiguous "standard." The standard obviously is only going to be employed to approve what organized medicine wants accepted. Since the insurance company has no incentive to argue, the current arrangement works well. It leaves only the chelation patients and their doctors unhappy.

THE MEDICARE FAIR HEARING FOR DENIED REIMBURSEMENT

As was pointed out in Chapter 5, any Medicare recipient who is denied reimbursement for bills entailed as a result of taking chelation therapy may request that his claim be heard by a Medicare Hearing Officer.

You make your appeal for a "Fair Hearing" at a local office of the Social Security Administration. To be compensated for your medical care payments, your hearing officer will look for treatment costs that were "reasonable and necessary." Show your physical need for chelation therapy, using X-rays, noninvasive test results, laboratory test results, witnesses (including your chelation physician), and anything or anyone else to verify the validity of your claim. Such exhibits should give the hearing officer virtually no alternative other than to pay your claim.

A patient given chelation therapy by Paul V. Wynn, D.O., of Albuquerque, New Mexico, won the first Medicare Fair Hearing for chelation therapy under Part B in September 1981. The Fair Hearing Officer deemed it "reasonable and necessary" that Dr. Wynn's patient have chelation treatment. He ruled that each intravenous EDTA injection is worth a minimum of $51.50 and paid this amount times the number of injections. The Fair Hearing Officer also ordered the Albuquerque Medicare carrier to pay the costs of all laboratory and noninvasive tests the patient had performed.

Unfortunately, two weeks after this apparent victory, Medicare rules regarding drugs were apparently changed again, effective October 1, 1981. A new bill, approved by Congress, says that no drugs may be paid by Medicare "unless they are effective for all categories of use." This statement seems highly restrictive. No one in government or in medicine as yet understands exactly what the new phraseology means, but it is not good news.

Possible negative interpretations prevail. For example, drugs proven effective for sore throat may be only "possibly" advantageous against pneumonia. Technically, the new law states that such a drug can't be paid for at all because it isn't beneficial in all categories of use (for sore throat *and* pneumonia). Accepting this negative interpretation, it could develop that Medicare will be paying for very few drugs. This conforms closely to the Reagan Administration's desire for cost cutting of social services, especially in Social Security's health care, although at the time of writing this book, the exact implementation of these new regulations is still unclear.

A new court ruling won by H. Rays Evers, M.D., of Cottonwood, Alabama, against the FDA says that a physician can utilize any drug for any purpose. This recent court decision may appear to conflict head-on with the new congressional law for Medicare payment of drugs. Consequently, there is now no clear-cut Medicare payment policy regarding chelation therapy by the Social Security Administration. Whatever appears to be the situation one day is clearly open to change. The change in favor of chelation therapy could come about with massive public interest and pressure on your elected representatives in Washington.

The current regulations create a confusing situation. It could possibly develop that those Medicare recipients seeking reimbursement for chelation therapy rendered before October 1, 1981, may collect on their bills and those who receive treatments after this date could be denied, at least until public pressure is brought to bear. Such a ridiculous circumstance requires a more uniform policy by the government. What we need is a massive outcry on the part of the public made in writing and by phone and telegrams to their elected federal representatives.

All of you who are reading this book who have potential problems from hardening of the arteries (which means virtually everyone living on a Western diet), should demand that research on chelation therapy be carried forward by the National Institutes of Health (NIH). Funding should also be made available for other medical research facilities to do controlled studies that private-practice AAMP physicians cannot possibly do. Furthermore, a "current consensus development conference" should be held under the auspices of NIH, with the input of every physician who has actually worked with EDTA sought and the entire proceedings recorded and made available to all interested parties. Tell your congressman that you want chelation research undertaken as soon as possible so that needless suffering and vascular surgery can be avoided. Reimbursement should be demanded from Medicare and the various private health insurance carriers, at least until bypass surgery is made to undergo the same rigorously controlled single-blind studies that opponents of chelation demand chelation patients undergo.

We should also have our congressional representatives request the Office of Technology Assessment (OTA) of the U.S. Congress to provide an objective analysis of this entire chelation question. The Office of Technology Assessment's function is to explore complex issues involving science for Congress. It does not advocate policies but sorts out the facts. Their September 1978 report assessing the efficacy and safety of medical technologies indicates that they have experience in dealing with this type of complex issue; a published report by OTA may help lead to new legislation to solve some of the problem of new uses for old drugs such as EDTA.

The AMA Drug Evaluation book has never contacted AAMP for EDTA information. The AMA appears hopelessly bias and outdated as a source of information to help the practicing physician learn about chelation therapy. Therefore, an NIH-sponsored Consensus Development Conference on Chelation Therapy could be useful. It must be an open and nonlitigious meeting of leaders in the field where all the important new information could come out. Pro and con opinions would be rendered, with no one getting sued, embarrassed, or insulted. The Consensus Development Conference could best be designated for fact finding and uncovering the truth about EDTA chelation therapy and could represent the first positive step forward. Then, some fair standard for judging the effectiveness of chelation treatment in vascular disease must be found, such as comparing a five-year follow-up of chelation patients to those on other medical treatments.

The following form is the type of letter you should write (consider using your own wording when writing your legislator):

> Dear Congressman (or Senator):
>
> I have read about EDTA Chelation Therapy to prevent or reverse hardening of the arteries. It's astounding to me that a treatment with such great merit has been overlooked and is receiving no official interest or research. Why aren't hearings being held by a suitable governmental body? Why don't we have a Consensus Development Conference and get the OTA to also look into the whole problem? Why is NIH not doing any research in this area? Why does only the AMA "standard" apply to what gets paid by my health insurance? You must help do something to bring this information to the public so that chelation therapy can be closely scrutinized and possibly adopted by American physicians as standard treatment.
>
> It's an outrage that Americans are dying from heart disease, diabetes, stroke, and other degenerative diseases that are manifestations of hardening of the arteries when chelation therapy could prevent or reverse the process. I want something done to make this treatment available for myself and my loved ones. Please take action at once!
>
> Thank you for representing my interest in this way.
>
> Your constituent,

The combined force of an irate public and the conclusions of a Consensus Development Conference could produce action from the usually ineffectual FDA. Simply by officially informing physicians that EDTA is a calcium chelator and thus may have calcium blocker or calcium antagonist actions, and that it has been extensively employed in medical practice for over twenty-five years, the FDA could help many physicians decide that this treatment might warrant their closer attention. They could then compare the literature on it to all the other new calcium blockers now beginning to be approved by FDA. Any such open, unbiased comparison could finally help many physicians to learn enough about chelation to begin to use it in their practices.

This simple government action to categorize EDTA as a calcium chelator and potential calcium blocker could be virtually automatic by the FDA, because the first words in the package insert that the FDA has approved are:

> EDTA forms chelates with . . . calcium. Because of its affinity for calcium, EDTA will lower the serum calcium level during intravenous infusion. Slow infusion may cause mobilization of extra-circulatory calcium stores . . . which is excreted in the urine. It is indicated . . . for hypercalcemia . . . and the control of ventricular arrhythmias and heart block associated with digitalis toxicity.

Yet, without a strong, organized public demand for such simple action, nothing will occur, because our drug-approval system was not set up for a drug company to have no interest in promoting its own drugs! FDA is a police-type agency, set up to prevent drug companies from making excessive claims, not to evaluate a therapy on its own.

The buzz words in Washington, D.C., today are *cost-effectiveness, cost-benefit*, and *competition*. Chelation therapy not only saves lives and limbs, it also matches the needs identified by those buzz words. We deserve access to this treatment.

CHAPTER ELEVEN

The Political Economics of Chelation Therapy

As you no doubt have gathered by now, of all the effective procedures applied in the United States to counteract cardiovascular disease, chelation therapy may be among the most uncommon. Orthodox physicians seem not to have read their own medical literature, in which chelation therapy is well described. Too often, upon being asked about the procedure by a patient alarmed by his or her own symptoms of disease, the doctor may admit to ignorance and say: "Bring me some scientific papers on the subject." Supposedly the doctor is "too busy" to do any literature search on his own, and his poor patient is forced into scrounging inadequately among vague information sources. This happens frequently.

Remaining ignorant of chelation therapy, such a "busy" physician will probably express suspicion of the treatment, as well. The doctor finds it easier to just throw out the statement "It's dangerous!" to cover a blow to the ego by one's personal lack of medical knowledge.

Worse than physicians' ignorance, we learned in Chapter 10 how third-party insurance carriers and organized medicine have helped to keep chelation therapy unavailable to Americans. If its cost was reimbursed to the policy holder, chelation therapy in general use would have a massive economic impact on the payout policies of health insurance companies.

And a marked income reduction would fall upon invasive cardiovascular specialists.

The financial issue is a strong one. In 1980, $54 billion was paid out in health insurance benefits. In 1981, $75 billion was reimbursed or paid over to health professionals on behalf of policy holders. Higher payouts are projected in 1982 and later years, with a 15 percent annual health insurance premium increase. Political economics is bound to infiltrate the medical care system of the United States. We are, after all, a capitalistic society with the profit motive our driving force.

Is there any wonder that special-interest groups prevail in medicine as in most other businesses? Don't be fooled by idealistic goals broadcast or printed by the media. When a cardiovascular specialist finishes four years of medical school, five years of hospital residency training, and two years more of subspecialty training, with all the effort that goes with getting his or her credentials, such as 75-hour work weeks, triple shifts, low wages, physical stress, emotional fatigue, and much more, he or she wants a big piece of the financial pie. And it's deserved! The problem is that the patient—you or I—provide the pie.

So, special-interest groups, usually the cardiovascular specialists, have tried to eliminate chelation treatment for hardening of the arteries from the American scene. They've been unable to accomplish their mission because the therapy is virtually the only antidote for heavy metal toxicity. Anytime a conscientious doctor becomes concerned with lead poisoning or some other form of heavy metal toxicity present in his or her patient, the organized medical community is in complete agreement on the method of treatment. "Fine, go ahead and use chelation therapy!" is the orthodox admonition. This indicates that physicians practicing traditional medicine believe the chelation procedure is sound, effective, safe, and they don't worry about its so-called toxicity. But as soon as a doctor wants to apply the same treatment for symptom reversal in heart attacks, he becomes a villain in the eyes of his orthodox colleagues.

The reason for such therapeutic hypocrisy is simply political economics. Few physicians in mainstream medicine care much about lead poisoning. Recognition and treatment of this subtle health problem has not achieved the popularity that it deserves. But whenever a chelating physician turns his therapeutic attention to poor circulation, he steps into another subspecialist's domain and reduces some financial remuneration from patients for that particular subspecialist. The subspecialist becomes a victim of his own training and medical beliefs, too, at the expense of the chelator.

"Hey, stay over there in occupational medicine," say the cardiologists, heart surgeons, and other cardiovascular specialists. "Treat all the cadmium, aluminum, mercury, and copper poisoning you want. That's OK!

But don't you infringe on my income from my bypass operations, drug prescriptions, catheter studies, and hospitalizations in our intensive care wards."

Obviously, a large group of chelation opponents evolves from the collective efforts of vascular surgeons, heart surgeons, chest surgeons, anesthesiologists, the administrators of hospitals with bypass surgical facilities, cardiologists, angiologists, internists, other physicians, and self-serving politicians from the American Medical Association. These specialists collectively claim that the EDTA chelating agent is dangerous and unproven. They point to the lack of double-blind research studies. Yet these same doctors permit and even encourage unproven bypass surgery to go on, with a known high death rate.

For example, William Stanford, M.D., chief heart surgeon at the Wilford Hall Medical Center, Lackland Air Force Base in San Antonio, Texas, has a 1976–1977 patient mortality rate of 43 percent. No action had been taken against him, however, until the deaths of seven children who underwent bypass surgery at the hands of this cardiovascular specialist who caused the anesthesiologists working with Dr. Stanford to rebel. The hospital's response was to curtail heart surgery at Wilford Hall temporarily and to transfer Dr. Stanford to civilian duty at St. Luke's Hospital in Milwaukee. A malpractice suite stemming from a Milwaukee case involving Dr. Stanford was judged against the United States Government.[1,2]

The Technology Assessment Act of 1972 called for the creation of a professionalized center for studying the potentialities and impacts of emerging technologies. The U.S. Congress established the Office of Technology Assessment (OTA), designed to provide new depth and expertise to the work of committees of the House and Senate in dealing with technological controversy and risk, and complementing the efforts of the Congressional Research Service and the General Accounting Office. Today, OTA stands on its own feet as an effective and respected center of scientific and technological policy analysis. Its agenda is determined primarily by the questions put to it by committees and subcomittees charged with legislative responsibilities.[3]

In 1978, Congress asked OTA to validate conventional medical practices in this country. OTA's report assessing the *Efficacy and Safety of Medical Technologies* concluded that "only 10–20 percent of all procedures currently used in medical practices have been shown to be efficacious by controlled trial." Accordingly, the request for controlled chelation investigations is irrelevant; other types of clinical evaluations should be allowed. When opponents of chelation therapy ask for double-blind studies as proof that the procedure works, they disregard the fact that validation does not exist for 85 percent of what they use in conventional medical practices. What is now being approved for health insurance reimbursement is clearly a political and economic decision—not a scientific one.

Since the practice of medicine is an art, you have to leave it open for the public to have access to the various professional artists. One patient may not do well in the hands of a neurosurgeon and actually might need the services of a chiropractor. Another may not get results from a cardiovascular surgeon and really need help from a chelating physician. These doctors are artisans with individual skills.

THE FLORIDA SUPREME COURT RULES FOR CHELATION THERAPY

Mainstream medicine may attempt to prosecute doctors who use chelation therapy to reverse or prevent hardening of the arteries in their patients, but the chelating physicians stand on their legal right to do so.

Typical of actions won in the courts by members of the American Academy of Medical Preventics is the case of Robert J. Rogers, M.D., of Melbourne, Florida. The frustrated Florida Board of Medical Examiners had asked the State Supreme Court to review an appeals court ruling against them that allowed Dr. Rogers to continue to use chelation treatment for hardening of the arteries.

A Florida First District Court of Appeal on January 11, 1979, reversed an order by the medical examiners placing Dr. Rogers on probation for a year and directing him to stop using chelation therapy. The hearing officer for the Florida Board of Medical Examiners had previously described the controversial therapy as "quackery under the guise of scientific medicine." The three district judges not only disagreed with the board's characterization of Dr. Rogers as a quack, but saw in him a scientific and medical innovator comparable to Freud, Pasteur, and Copernicus, all of whom were derided by their contemporaries. The judges made this comparison in their written decision.

Michael Schwartz, the medical board's lawyer, appealed the decision on two grounds. He said that the case is in conflict with prior court rulings and it involves a constitutional question. Dr. Rogers's lawyer, Andrew Graham of Cocoa, Florida, agreed that the case was a constitutional issue but not on prior court rulings because of dissimilarities with other cases.

The beleaguered Florida Board of Medical Examiners lost its appeal. And the opinion has far-reaching implications. It has become a main step in bringing down the barriers to chelation therapy and other innovative procedures not generally accepted by mainstream medicine. The appellate court ruled and the Florida Supreme Court has now upheld the ruling in favor of the intravenous EDTA treatment. Dr. Rogers's patients and all other patients receiving chelation therapy in the State of Florida have a constitutional right to receive the injections as long as they do no harm and involve no fraud.

The Brevard County Medical Society, which first initiated action against Dr. Rogers, reacted by demanding that the Florida Board of Medical Examiners fire its executive director, George Palmer, M.D., and also fire attorney Schwartz for allegedly botching the case. The furor raised by execution of those demands spurred the Florida State Legislature to take a closer look at what the medical board was doing and how medicine was being administered in the state.

The result has been legislation that strips this medical board of examiners and twenty-four similar professional boards of the power to investigate complaints. This authority is now transferred to the Florida Department of Professional and Occupational Regulation, which, in turn, makes recommendations to all the boards.

The case of Dr. Robert Rogers clearly shows that physicians have the right to offer a treatment even if mainstream medicine has not fully adopted it. When a significant minority of physicians believe in a particular therapy, their individual licensure permits them to prescribe the remedy to benefit their patients. They take full responsibility for the treatment's effects—good or bad.

Dr. Rogers had brought in chelating colleagues as expert medical witnesses to testify when the Brevard County Medical Society first attacked him. They testified from practical experience derived from the use of intravenous EDTA injections for their patients. The experts presented patient histories and described cases of improved blood flow through formerly clogged arteries. Their testimony had no effect whatever on the medical society, which immediately ordered Dr. Rogers to stop using chelation. The next day the society turned him over to the State Board of Medical Examiners for discipline.

His patients were not allowed to testify as to their remarkable recoveries from degenerative disease when Rogers appeared in front of the medical board. The medical board's witnesses, opposing chelation therapy, did not have any practical medical experience from treating patients with chelation to draw upon. The plaintiff's "experts" relied merely upon their brief review of medical textbooks and made broad assumptions against chelation based on theory and bias.

This is the typical situation whenever chelation therapy has been challenged legally—the opponents are never adequately qualified experts. They have not used EDTA personally in the treatment of vascular disease.

None of the opponents have ever taken specific in-depth training in chelation therapy for vascular disease such as that provided by the American Academy of Medical Preventics. In the six years I have been reporting on and researching this subject, I have never seen a witness against chelation who had enough knowledge about the treatment to even be able to attempt to pass the written and oral examinations offered by the American Academy of Medical Preventics. The situation is as clearly absurd as it would be for a psychiatrist to review a urologist's surgical practice, or a pediatric radiologist to judge a gerontologist's work. Even so, the courts are afraid to tell the big medical chief from the local university that he is not adequately qualified to testify because he lacks firsthand knowledge of the therapy. The courts generally fail to point out to the jury that, in fact, the local "expert" against chelation has a strong vested economic interest in status quo medicine and is significantly threatened by the new medicine—typified by resistance to such things as DMSO for arthritis, Dilantin when used for pain, headaches, or depression, and chelation therapy when used for hardening of the arteries.

The administration of chelation therapy has indeed become a new subspecialty in medicine. No medical consultant should render an "expert" opinion about the treatment in a legal proceeding unless he can show that he is fully informed in the field and is able to meet the standards set forth by the American Academy of Medical Preventics.

Even though AAMP physicians had not conducted what the opponents would accept as adequately controlled double-blind clinical evaluations, such as are ordinarily required by the Food and Drug Administration when a pharmaceutical company is trying to get a new drug approved, the chelation proponents in the Rogers case displayed such extensive clinical and personal experience that Dr. Rogers's use of chelation therapy had to be legally approved by the State of Florida. Since then, four small but better-controlled clinical studies have since been carried out, using extensive noninvasive tests such as radioisotope coronary stress studies. Additionally, the AAMP has on record over 18,000 before-and-after complete thermographic studies of patients who, prior to taking chelation therapy, had circulatory impairments. Their conditions were carefully documented with these thermographic studies, which can be compared to National Bureau of Standards Controls and are precisely calibrated to an accuracy of .01 degree centigrade.

These circulatory impairments involved peripheral and/or cerebral studies performed by P. P. Hoekstra, M.S., Director of Clinical Thermography of Therma-Scan, Inc., St. Clair Shores, Michigan. He is currently working to have all of this information computerized. Hoekstra's goal is to make it possible to predict how many treatments will produce how much improvement in a patient with a given amount of vascular impairment. Thermographic studies are one of the easiest ways to screen the population to identify potential stroke-risk patients. They then may need additional confirming noninvasive studies, such as radioisotope scans. The patients could then be improved with adequate chelation therapy, and restored to a higher level of functioning, in many cases with actual reversal of senility and other cardiovascular conditions.

ILLEGAL RESTRAINT OF TRADE BY THE AMA

It is particularly inappropriate for any individual, institution, or agency to turn to the American Medical Association for an adequate and unbiased opinion about chelation therapy. The AMA was legally determined by the Federal Trade Commission to be "guilty of conspiracy" in enforcing its "ethical" standards. These standards constitute unfair monopolistic methods of competition and amount to an illegal restraint of trade in violation of Section 5 of the Federal Trade Commission Act. Such illegal AMA guidelines have impeded progress in new approaches to health care. This case, known as the Barnes decision, was reviewed this year by the Supreme Court of the United States, and, of course, the AMA was desperately hoping to have the Barnes decision overturned. It was not! After all, organized medicine argued it was only trying to protect the public from bad medical practices.

This argument sounds reasonable until you really evaluate the true "cost" we pay the AMA for this supposed "protection" from unethical or incompetent doctors. It now appears clear that under the guise of protecting us, organized medicine can keep major advances in treatment from becoming accepted for years, while thousands needlessly suffer, or are given outmoded or inadequate treatment.

Incidentally, the AMA's 152,000 full dues-paying members in 1980 represented only about one-third of the nation's medical doctors. Today the AMA must compete with more than 200 other specialty groups that seek physician membership. It doesn't count as the true voice of American medicine anymore.[4] And it really never was.

A Federal appeals court has substantial evidence to support its finding that the ethical standards of the AMA were used illegally to restrict competition among doctors as charged by the Federal Trade Commission. In an October 1980 ruling, the Second U.S. Circuit Court of Appeals upheld the FTC order by Judge Barnes that had directed the AMA to stop enforcing so-called ethical restraints on physicians. The FTC, in its landmark 1979 ruling, found that there is ". . . a substantial body of formal and informal actions by the AMA that have the effect of enhancing the economic positions of members of the AMA. The end result has been the placement of a formidable impediment to competition in the delivery of health care services by physicians in private practice. The costs to the public in terms of less expensive or even, perhaps, more improved forms of medical service are great.

". . . AMA's ethical structures were motivated by economic objectives rather than by a need to maintain professionalism among physicians."[5]

The "Antitrust and the Health Professions" report—another study—prepared by Michael Pollard and Robert Leibenluft of the FTC's planning office concludes that physicians tread a fine line when they attempt to regulate themselves through their associations by sometimes breaking the restraint-of-trade laws. The authors note that professional societies "wield tremendous power" through their accreditation and certification programs. Medical associations are in legal limbo if they attempt to act as quasi-licensing bodies, as the AMA had done in making judgments on chelation therapy. The imposition of disciplinary measures abounds with potential abuses. Such measures "can be used to block innovative and experimental approaches to health-care delivery by non-physicians, alternative facilities, or traditional practitioners offering new kinds of treatment," the report says. "Expulsion from a medical society also can be used as a signal to other physicians that nonconforming practitioners should be boycotted."[6]

There are many chelating doctors, in a number of states, who are being hassled today in one way or another for their use of alternative treatment methods. More than twenty-five licenses are being questioned by medical boards of examiners. It is not the intent of this book to mention them all or to go into the merits of the cases; a few examples will suffice.

Hellfried E. Sartori, M.D., of Rockville, Maryland, is currently in a major fight with his state licensing board. The board of medical examiners insists he may not use hair analysis. His peers were willing to let him do nutritional and chelation therapies if Dr. Sartori gave up the right to be a primary physician and agreed not to prescribe drugs. They apparently felt that they really do not have the power to deny Dr. Sartori the right to do chelation therapy since the Florida Supreme Court case of Dr. Rogers and the Federal Appellate decision in the case of Dr. H. Ray Evers have shown that government does not have the right to stop chelation therapy. However, the Maryland board apparently wanted everyone who saw Dr. Sartori to know that his was not the regular practice of medicine. Dr. Sartori explained to fellow AMA physicians that the concept of a limited license was considered. But this limitation became too cumbersome to work out by the board of examiners, since even chelation doctors occasionally have to use drugs and surgery—at least for a time.

Stanley Olsztyn, M.D., of Scottsdale, Arizona, has received an order from his Arizona State Board of Medical Examiners to stop giving his patients chelation therapy. Inasmuch as the use of EDTA was banned in Arizona by this same board several years ago, Dr. Olsztyn chelates merely with citric acid, he said. Even this mild organic acid modality is now ordered banned in Arizona. The special-interest group that concerns itself with angiography and cardiovascular surgery seems to find that chelation therapy in any form furnishes too much competition in the financial battle for patient dollars.

THE DEFENSIVE MEDICAL PRACTICES OF CHELATING PHYSICIANS

It is common practice today for the State Board of Medical Examiners in many states to send phony "patients" into certain physicians' offices, feigning illness to see if the doctors are really practicing good medicine. Each state is ready to take disciplinary action against any doctor if he fails to do a "good-faith" job. He or she must take an adequate history, do a physical examination, and obtain appropriate tests before ordering any therapy, particularly a controversial one like chelation treatment. Because of their strong belief in chelation therapy, some doctors have been tempted to simply run chelation "mills" and give the therapy to anyone who can afford it, without first adequately checking to be certain that patients really have occlusive vascular disease by ascertaining the exact status of their circulatory system. The AAMP protocol requires them to use the new noninvasive tests in conjunction with a careful history and physical examination. The possibility of medical discipline helps to prevent any potential abuse of the treatment. At the same time, it makes chelation therapy more expensive for those who need it.

If you wonder why the costs of obtaining chelation therapy are so high, part of the problem is that chelation doctors are constantly on the defensive. Extensive noninvasive vascular and laboratory tests are performed to help protect the doctor against this program of possible entrapment by phony patients or shills, as well as from groundless malpractice suits. The malpractice risk is excessively high because most nonchelating physicians routinely provide incorrect information about chelation therapy and its potential benefits and/or toxicity to patients who inquire about it. Most nonchelating doctors do not understand how chelation works or recognize that it may take several months to produce its maximum benefits. Overanxious patients are easily persuaded to discontinue chelation therapy prematurely and are often convinced by these generally uninformed and frequently rather biased doctors that they have completely wasted their money.

Many chelating doctors have lost their malpractice insurance over exactly this kind of legal harassment. Thus, today's chelating physicians practice defensive medicine, which includes doing all the latest tests to prove the exact circulatory status before and after chelation on every patient. This documentation, employing fully approved noninvasive tests that are completely recognized by Medicare and other insurance companies, constitutes objective evidence of the need for treatment. The evidence can be shown to a jury, if the doctor is challenged. There is nothing wrong with these noninvasive vascular tests, other than that their added expense may make chelation therapy financially unavailable to some patients.

The many chelation doctors today who are forced to practice without malpractice insurance are in an unfortunate position. Most physicians would prefer not to leave patients without this important protection, since potentially serious errors can occur in any practice, and meritorious claims do exist. Still, malpractice insurance is often simply not available at any price to chelating physicians, partly because the malpractice insurance companies are concerned regarding their liability because of today's unfriendly medical attitude toward chelation.

Malpractice insurance companies recognize that usually no one thinks about suing the doctor or has little chance of winning if a patient dies after he has undergone "standard" treatment, even if it does nothing to help the patient. For example, there are very few suits filed on behalf of approximately 7500 who have died and many thousands more who suffered serious complications in the past one year alone, as a direct result of the essentially unproven and excessively performed "standard" bypass surgeries.

Death or disability are still generally accepted as "God's will" or "fate" when it follows so-called "standard" therapy; yet, when alternative medicine fails, many will immediately blame treatment failure, and it is often assumed by everyone that the patient might have been better off had he stayed with "standard" treatment, even though there are no facts to support that concept.

Unfortunately, if a patient claims, right or wrong, that chelation treatment has somehow harmed him, or if he simply claims that it has failed to help him get well, there always are plenty of eager attorneys around who have unlimited access to hostile "expert" medical witnesses. These medical witnesses usually have no idea what EDTA chelation therapy is, yet they testify against chelation doctors who are truly specialists.

Malpractice companies, therefore, have some legitimate reason to be concerned about insuring a chelating physician. Furthermore, since the medical literature shows that only approximately 80 percent or so of chelated patients will be demonstrably helped after the first twenty to thirty chelation treatments; not everyone is completely satisfied after taking chelating therapy—especially if they had to pay for it entirely from their life savings. This source of potential ill-feeling is largely avoided by physicians' "standard" therapies, because then the patient usually gets at least 80 percent or more reimbursed by his health insurance. Since the cost didn't really come out of his pocket directly, patients are less inclined to sue or be upset in case of treatment failure, and will often either live with their health problem or go "doctor shopping" for another type of treatment to try.

Legal proceedings are impeding the use of these new alternative medical practices in our country, and thus they are helping to retard medical progress. Many doctors, seeing the legal risk, are simply unwilling to expose their families to these potentially serious economic hazards, just to try and help a patient by offering a new or controversial therapy. This situation has led to the "closet chelator" in the United States, who only treats himself or his close relatives and/or friends and routinely denies that he does chelation therapy to anyone who asks.

Recently a large settlement was made out of court on alleged kidney toxicity case in Texas involving EDTA. The doctor's malpractice insurance company was afraid of the potential of an even larger settlement from a sympathetic jury who might be emotionally swayed after hearing the prosecution's "expert" medical witnesses testify how bad they believe chelation therapy is and how dangerous they believe it is. The patient really had a major kidney ailment. However, there was no evidence to document that his kidney problem had any connection whatsoever to the chelation therapy he had received earlier. In fact, there was excellent evidence indicating that the kidney problem the patient later developed could not possibly be related to his prior chelation therapy.

This case was not brought to trial and a particular chelation doctor has had her malpractice insurance canceled. In the future she may be less willing to try and help her patients by going the "extra mile" for them. She may become much more cautious and less innovative, or possibly even leave medical practice. In this way, many patients will potentially suffer, because of one unjustified legal action that led to an out-of-court settlement.

All too often the insurance companies are afraid to let the chelation doctor go to court and defend himself against what are often ridiculous charges. The insurance companies point out that the jury is always sympathetic and tend to award money to anyone who appears to be seriously ill or in pain, no matter who was at fault. They fear the jury cannot understand that the chelation doctors are correct about the validity of chelation therapy, and that the distinguished-sounding and well-paid medical "experts" who appear to testify against chelation therapy are completely wrong.

Some experts able to testify for a chelation doctor who is being unjustifiably sued over an issue such as alleged renal toxicity may refuse to do so because it could have an adverse effect on their own professional careers. There are many powerful interests that want chelation stopped.

Other experts will testify on behalf of EDTA only for heavy metal toxicity and not when it is used in vascular disease. This is possible because even a chelation "expert" could study all of his life, know a great deal about EDTA, and still have no knowledge about the effectiveness of chelation therapy in vascular disease. Many doctors who use EDTA for lead poisoning seem to feel threatened by doctors who dare to use it for a new use such as vascular disease. They seem to ignore the evidence rapidly accumulating from around the world that proves chelation therapy is highly effective.

In the middle of this complex and highly charged situation, we find that state licensing boards have independently decided to crack down on bad or incompetent doctors, often because of strong political pressure. The state licensing boards are obviously very new to this function of trying to clean up medicine. Until now, doctors have generally been left alone no matter how badly they practiced unless they were continually drunk or were abusing hard drugs, or they had sexually molested their patients. Now medical costs are soaring out of sight, and the need to improve medicine has become critical.

The investigations in most states tend instead to be directed toward the more controversial physicians—such as those using alternative health practices. Since these doctors usually do not have the political "muscle" of organized medicine, which always vigorously defends "standard" treatment, prosecutors with the aid of physician politicians are able to convict these doctors quite easily and make examples of them. This helps keep in line the rest of the practitioners who may have been thinking about becoming involved in alternative medicine.

To help build their case against holistic doctors in court, the state medical licensing authorities may employ phony patients or shills who usually pretend to have a nonexistent illness when they see the physician. These shills may be employed by the federal government, state government, health insurance companies, or private industry such as the newspaper or electronic media.

It is very difficult to refute accusations made by such medical shills or "plants," or to override patients' biased testimony. These people are easily convinced by the investigators that they are helping the state get rid of a dangerous "quack" doctor. Somehow these people often recall only what they want to remember. Often this testimony may be severely distorted so that it serves to discredit the physician. The good that the doctor has done them seems easily forgotten by many witnesses in the excitement of the trial. Still, tape recorders can be an effective defensive tactic against false testimony and help the physician in his defense. Their routine use is now being increasingly undertaken by chelation physicians.

Chelating physicians have been advised to invest a few dollars a day to keep a cassette tape recorder going during all of the doctor-patient contacts from his daily professional life. These sessions are recorded on long-playing tapes and are kept in a file. Furthermore, individual physicians are urged by AAMP legal counsel to request their patients to bring their own tape recorders with them to every office visit. This way the patient can listen again to the doctor's instructions and his family can more easily cooperate and help him adhere to his new lifestyle, often achieving better health for the entire family. This type of combined education and defensive action should help discourage needless litigation over professional malpractice and should also help keep testimony more honest.

"The whole world is laughing at medicine as it is practiced in the United States," says James George Defares, M.D., Ph.D., Professor of Research Medicine at the University of Leiden in Holland, who is representative of doctors outside this country in his attitude toward American physicians. "The way American doctors function in their daily delivery of health care is equivalent to living in a police state. And the gestapo enforcing the rules of medical practice are two agencies, with one backing up the other. The first despotic group is the American Medical Association, maintained by the moneyed physician-politicians. Their supporting arm is the Food and Drug Administration, representing the interests of the pharmaceutical industry. I don't hold either group in high esteem, and neither do most of my colleagues in Europe."

Functioning under stressful conditions, American physicians providing the chelation answer to hardening of the arteries are continuing to learn to keep more careful patient records. They aim to demonstrate objective proof of increased circulation in their patients after receiving the treatment. The patients remain their own controls, and their laboratory tests tell the objective story.

They know that someday the results will finally come in from all the ongoing research activities. They will finally produce widespread general agreement on the big questions about chelation: (1) How many of your chelation patients actually have proven reversal of hardening of arteries? (2) How many patients go how many years without getting a heart attack or stroke, or having to go to the hospital for illness? (3) How many return to work? (4) How many really can be proven to live longer than patients receiving traditional forms of care? When this data is finally in and broadly disseminated, then there will be a discontinuance of the harassment and persecution by their more traditionalist colleagues.

In the meantime, approximately one million people a year get heart attacks and over 600,000 are needlessly dying from a treatable and preventable disease. Recent figures released from NIH indicate that a dramatic 20 percent decrease in heart attacks and as much as a 40 percent decrease in strokes have occurred in the United States over the last few years. This shows that public awareness of the dangers of smoking and high-fat diets, the need for exercise, the widespread use of vitamins and supplements, and the increasing use of newer diagnostic vascular testing and therapies such as chelation therapy are beginning to have an overall beneficial effect on the health of this nation. It provides strong evidence that the problem is surmountable—with a multifaceted approach as advocated in this book. Chelation therapy can play a central role in helping to stop this senseless loss of life. Your passing this book on to a friend will help hasten that day.

THE FDA-PHARMACEUTICAL INDUSTRY SPECIAL-INTEREST GROUP

The most frustrating form of harassment for chelating physicians, of course, comes from one's own government. A government agency such as the FDA is designated to provide health protection to everyone. But U.S. Representative James H. Scheuer, Democrat of New York, charges, "The Food and Drug Administration is contributing to the needless suffering and death of thousands of Americans because it is denying them the life-enhancing and life-saving drugs that are available abroad far sooner than they are here. . . . It takes substantially longer to get a new drug approved in America because of FDA administrative inefficiencies and regulatory excess."[7]

The issue of an "orphan drug" such as ethylene diamine tetraacetic acid, one having no pharmaceutical company sponsor because it can't be protected by a patent, is enormously complicated. It touches on return for investment by a pharmaceutical company, patent policy of the U.S. Patent Office, drug regulation by the Food and Drug Administration, testing criteria of the scientific method recognized by medical researchers, and bureaucratic delays in approving old drugs for uses that are new or new drugs for diseases that are different but that have similar symptoms, meaning that drugs may be found that will help some but not all of its victims. EDTA is caught in this morass of political economics generated by impediments put out by the FDA.

The most recent expression of the incredible magnitude of this fatal flaw in our system appeared on the editorial page of the *Wall Street Journal* for Monday, November 1, 1981, under "Review and Outlook," page 26, under the title "100,000 Killed." This editorial discusses the scandal regarding how the FDA has obstructed the use of one drug—propanolol (Inderal), which took twenty years (after this class of drug first appeared) to receive FDA approval as a preventive for second heart attacks. The *Wall Street Journal* goes on to ask whether we should even have FDA any longer and states that no one has added up the lives that would have been prolonged in the absence of regulatory delays caused by the Agency. "We are . . . convinced . . . that they would far outweigh the number of lives lost or damaged if the responsibility for safety were merely returned to the drug makers and doctors. It should be kept in mind that drug makers and doctors also have a vital interest in drug safety for their own protection. It is clear that FDA bureaucrats will never take any risks they can avoid—they have nothing to gain by approving an effective drug and everything to lose if they make a mistake. This kind of approach guarantees a huge loss of life." The newspaper asks, how much longer should it be allowed to prevail? This statement was made by the *Wall Street Journal* without the staff having any knowledge of the millions of lives and billions of dollars that could have been saved if EDTA's use in vascular disease had not been obstructed by the FDA, with the help of all whose vested interests benefit from this obstruction.

When Arthur Hull Hayes, Jr., M.D., former Chief of the Division of Clinical Pharmacology at Pennsylvania State University College of Medicine, stepped into the Commissionership of the Food and Drug Administration, he took an oath of office. His responsibility is clear—to protect consumers against impure and unsafe foods, drugs, and cosmetics. The seven to ten years and nearly $30 million it takes today to get a new drug approved deprives consumers of many beneficial medicines. Dr. Hayes must address this problem by completely altering the agency's drug-approval process. Even better, President Reagan should simply remove the responsibility for total control of drugs from the FDA by completely changing the laws under which drugs are approved for new uses today.

One clear way to accomplish the task is to adopt the findings of a special panel, commissioned by the federal government, which had conducted a two-year study of the FDA. It reached the same conclusion in its final report of May 2, 1977, as had many outside critics. The panel, headed by New York University law professor Norman Dorsen, advised that the agency's ties with the drug industry it regulates are so cozy and so secretive that they undermine confidence in the objectivity of its decisions. Informal meetings and telephone conversations between pharmaceutical company representatives and FDA officials form a cloud of suspicion over the way the agency does business.

The Dorsen panel recommended that procedures be adopted to reflect the fact that no drug is absolutely safe and always effective. It suggested a new legal standard "to reflect the fact that assessing the value and ultimate approvability of a new drug entails weighing its risks against its over-all benefits."[8]

This report is now nearly five years old, and none of its recommendations have been put into practice. General bureaucratic inertia seems to be the reason. The same old secrecy shrouds the FDA regulation of drugs. The same old discrimination is being carried out by the agency against alternative medical thinking and innovative practices. The FDA is responsible for repeatedly impeding progress in American medicine and therefore causes substantial harm to the public interest and a waste of taxpayers' dollars.

For example, the FDA sued to enjoin mislabeling of EDTA by H. Ray Evers, M.D., an individual physician doing business at the time as Ra-Mar Clinic in Montgomery, Alabama. The government brought suit, in 1978, in the United States District Court for the Middle District of Alabama, Robert E. Varner, Chief Judge, legally identified as 453 F. Supp. 1141, presiding. The FDA lost its case and then appealed.

The object of the FDA case against Dr. Evers was not his prescription of EDTA for use in the treatment of circulatory disorders. Instead, the government sought to challenge the physician's promotion and advertising of chelating drugs for that use. According to FDA attorneys, Dr. Evers "misbranded" the chelating agent when he publicly advocated it for an unapproved purpose without providing "adequate directions" for such use. In other words, Dr. Evers could use EDTA against hardening of the arteries, but he could not talk about his treatment of the condition with this medication, and if he did not talk about the manner and reason for the medication's use, he was misbranding the drug.

The court worked its way out of this *Catch-22* situation by ruling only on the narrowest of issues of whether Dr. Evers violated section 301(k) of the Federal Food, Drug, and Cosmetic Act, as amended August 1972. This section was intended to keep misbranded drugs out of the channels of interstate commerce at each stage of the distribution process from manufacturer to consumer. Paragraph (k) deals with the period during which a drug is "held for sale after shipment in interstate commerce." It prohibits the changing of labels.

The court of appeals ruled that the FDA failed in its case to establish a violation of section 301(k) of the Act. "We therefore affirm the judgment of the district court in favor of Dr. Evers. . . . the Act was intended to regulate the distribution of drugs in interstate commerce, not to restrain physicians from public advocacy of medical opinions not shared by the FDA. We believe, therefore, that a doctor who merely advocates to other doctors a lawful prescription drug for a use not approved by the FDA, and does not distribute that drug to other doctors, is not holding that drug for sale within the meaning of the statute and therefore is not in violation of section 301(k) of the Act."

The opponents of chelation therapy were again proved wrong. In his opinion against the FDA, Chief Judge Robert E. Varner wrote: "The requirement which the FDA seeks to impose is nonsensical."

The FDA has but one way to stop a physician from chelating. The governmental body can remove a drug from the market if it can show general harm to the public. It's clear that the FDA has been unable to take away EDTA chelation therapy, because it isn't harmful and is efficacious. The FDA tried desperately to prove that the public was hurt in some way but could not do so. It spend hundreds of thousands of dollars in investigatory costs and legal costs and hundreds of man-hours with medical and legal minds on the subject. None of its allegations of harm held up in court. The FDA had its legal day and lost this time around. Patients can now feel quite reassured that the FDA put its best effort forward and made no case against chelation therapy.

The drug manufacturers of EDTA are prohibited by law from telling anyone what chelation can do in vascular disease. Violating this law opens them to extremely serious penalties after prosecution. The only way the EDTA manufacturers might show the ligand's intravenous therapeutic benefits would be first to spend approximately $30 million over an eight-to ten-year period to complete practically impossible double-blind studies on human beings. The EDTA makers therefore aren't able to inform traditional physicians about chelation therapy.

The FDA has the absolute power to block all sales of any drugs such as EDTA to anyone, instantly, if the agency can prove substantial harm to the public. Since the opponents have provided only unproven allegations against the many known and widely reported benefits of the substance, the FDA has been unable to legally stop its sales, although the agency has put tremendous pressure on EDTA manufacturers.

In one form of pressure, the FDA has ordered all pharmaceutical supply companies not to ship EDTA to physicians unless the companies first ascertain that the drug is not going to be used for any purpose except the ones listed in the package insert. This outrageous and illegal FDA demand is tantamount to having you ask someone purchasing a bottle of vitamin C or aspirin what he or she plans to use it for. Not getting the right answer, you are then forbidden to sell it to him. Such a situation matches exactly the virtual police state Dr. Defares describes, in which American medicine is practiced. It affects, in fact, everything that doctors do and is a violation of their constitutional rights as well as their state licensure rights as physicians. Somebody must stand up to the FDA and fight back, or physicians' rights will soon disappear altogether.

Jim R. Critchlow of Cooper City, Florida, the former president of Pharma-Syst Company and now the president of Phyne Pharmaceuticals, has fought the FDA for years at great personal expense and sacrifice to keep EDTA available for his physician clients. The FDA has practically bankrupted him by dragging the man through the courts, using our own taxpayer money as its best weapon against private citizens. "If you don't like our actions," say the FDA officials with a knowing smirk, "sue us!" They know the high cost of litigation against the government.

Owing to the dedication of drug manufacturers like Jim Critchlow, who has spent his life savings for the cause of EDTA chelation therapy, the FDA's efforts to ban the sales of this life-saving medication have been unsuccessful. EDTA remains available for all doctors and their patients in the United States. The American Academy of Medical Preventics officially recognized Jim Critchlow at the Academy's November 1981 conference for his untiring efforts to keep the chelation movement alive.

Hundreds of other "freedom fighters" for retaining EDTA chelation therapy in this country should be recognized. They are supported by hundreds of thousands of grateful patients. All of them are people dedicated to the cause of cardiovascular health.

Today it is rare to find a chelating physician who has not given himself the treatment, as well as most of the members of his family. These health professionals take chelation therapy to enjoy optimal health and to live a longer life. The chelated doctor agrees that he or she experiences increased efficiency in the ability to work and think. For himself, his loved ones, and his patients, the intent of the chelating physician is to reach for body and mind functioning at the highest level of wellness whether or not any cardiovascular disease signs or symptoms are present.

If you have ever asked yourself if you could have these same effects of optimal wellness—preventing hardening of the arteries and rejuvenating your cardiovascular system—chelation therapy is the answer.

APPENDIX I

Current Availability of Chelation Therapy for You

Chelation therapy in the United States is administered openly by more than 350 health professionals, most belonging to the American College of Advancement in Medicine (ACAM, formerly the American Academy of Medical Preventics or AAMP). While the ACAM membership is larger than the listing here, not all the members administer chelation therapy and not all of the chelating physicians want their names to be listed.

An estimated 800 additional American doctors are providing the chelation treatment clandestinely and are not participating in the open effort to legitimize it among the establishment medical practitioners.

Of the ACAM chelating physicians, some permit their names to be published. They have made chelation therapy their medical subspecialty and have organized to struggle collectively against colleague criticisms of their practice.

Other chelating ACAM members (Fellows) keep a low profile by offering the treatment as a regular part of patient management but making no announcement of the fact.

Certain more timid chelating doctors are not ACAM Fellows and avoid the severe and stressful antichelating pressure by giving treatment only to friends, relatives, specially selected patients, and themselves. The nonmember chelators sometimes attend the Academy's semiannual scientific sessions, which reveal new advances in the chelation procedure.

When the following referral directory of physicians providing chelation therapy was compiled, some doctors had more than one office and therefore are listed more than once.

The doctor's degrees are listed after his or her name. Please note that M.D. means "Doctor of Medicine," D.O. means "Doctor of Osteopathy," Ph.D. means "Doctor of Philosophy" in a health science, and N.D. means "Doctor of Naturopathy." Since a doctor of philosophy is not permitted to administer human treatment except under the supervision of a licensed physician, and a doctor of naturopathy adheres to the discipline of treating diseases largely by employing natural agents and rejecting the use of drugs and medicines, a doctor of philosophy or a naturopath is not considered a participating physician of ACAM. Rather, he is a member of the American Institute of Medical Preventics (AIMP) and as such is listed here as offering chelation therapy. Any M.D. or D.O. who is licensed for the intravenous administration of drugs, provides chelation therapy to his patients, and agrees to follow the recommended treatment regimen described in the ACAM protocol is considered a "participating physician."

The author cannot be certain of the qualifications of individual practitioners. You should consult your own physician and the doctor you select to discuss your needs before beginning any treatment.

By state, here is the referral list of chelating health professionals belonging to or adhering to the principles of the American Academy of Medical Preventics and the American Institute of Medical Preventics.

Alabama

P. Gus J. Prosch Jr., MD
759 Valley Street
Birmingham, AL 35226
(205) 823-6180

Alaska

F. Russell Manuel, MD
4200 Lake Otis Blvd. #304
Anchorage, AK 99508
(907) 562-7070

Robert Rowen, MD
615 E. 82nd Avenue, Ste 300
Anchorage, AK 99518
(907) 344-7775

Paul G. Isaak, MD
Box 219
Soldotna, AK 99669
(907) 262-9341
RETIRED

Robert E. Martin, MD
P.O. Box 870710
Wasilla, AK 99687
(907) 376-5284

Arizona

Lloyd D. Armold, DO
4901 W. Bell Rd., Ste 2
Glendale, AZ 85308
(602) 939-8916

William W. Halcomb, DO
4323 E. Broadway
Suite 109
Mesa, AZ 85206

S.W. Meyer, DO
332 River Front Dr.
P.O. Box 1870
Parker, AZ 85344
(602) 669-8911

Terry S. Friedmann, MD
2701 E. Camelback Rd.
Suite 381
Phoenix, AZ 85016
(602) 381-0800

Stanley R. Olsztyn, MD
Whitton Place
3610 N. 44th St.
Suite 210
Phoenix, AZ 85018
(602) 954-0811

Gordon H. Josephs, DO
315 W. Goodwin St.
Prescott, AZ 86303
(602) 778-6169

Gordon H. Josephs, DO
7315 E. Evans
Scottsdale, AZ 85250
(602) 998-9232

Garry Gordon, MD
5535 S. Compass
Tempe, AZ 85283
(602) 838-2079

Arkansas

William Wright, MD
1 Mercy Drive, Ste 211
Hot Springs, AR 71913
(501) 624-3312

Melissa Tallaferro, MD
Cherry Street, P.O. Box 400
Leslie, AR 72645
(501) 447-2599

Norbert J. Becquet, MD
115 W. Sixth Street
Little Rock, AR 72201
(501) 375-4419

John L. Gustavus, MD
4721 E. Broadway
N. Little Rock, AR 72117
(501) 758-9350

Doty Murphy III, MD
812 Dorman
Springdale, AR 72764
(501) 756-3251

California

Ross B. Gordon, MD
405 Kains Avenue
Albany, CA 94706
(510) 526-3232

Ralph G. Selbly, MD
1311 Columbus St.
Bakerfield, CA 93305
(805) 873-1000

Carol A. Shamlin, MD
621 E. Campbell, Ste. 11A
Campbell, CA 95008
(408) 378-7970

Eva Jalkotzy, MD
156 Eaton Road, #E
Chico, CA 95926
(916) 893-3060

John P. Toth, MD
2299 Bacon St., #10
Concord, CA 94520
(510) 682-5660

Michael Rosenbaum, MD
45 San Clemente Drive
Suite B-130
Corte Madera, CA 94925
(415) 927-9450

James Privitera, MD
105 No. Grandview Ave.
Covina, CA 91723
(818) 966-1818

Charles K. Dahlgren, MD
1800 Sullivan Ave., #604
Daly City, CA 94015
(415) 756-2900

William J. Saccoman, MD
505 N. Mollison Ave.,# 103
El Cajon, CA 92021
(619) 440-3838

A. Leonard Klepp, MD
16311 Ventura Blvd., #725
Encino, CA 91436
(818) 981-5511

David J. Edwards, MD
360 S. Clovis Ave.
Fresno, CA 93727
(209) 251-5066

Bruce Halstead, MD
22807 Barton Road
Grand Terrace, CA 92324
NO REFERRALS

James J. Julian, MD
1654 Cahuenga Blvd.
Hollywood, CA 90028
(213) 487-5555

Joan Priestley, MD
7080 Hollywood Blvd.,
Suite 603
Hollywood, CA 90028
(213) 957-4217

Joan M. Resk, DO
18821 Delaware St., Ste 203
Huntington Bch., CA 92648
(714) 842-5591

Carolyn Albrecht, MD
10 Wolfe Grade
Kentfield, CA
RETIRED

Pierre Steiner, MD
1550 Via Corona
La Jolla, CA 92037
NO REFERRALS

David A. Steenblock, DO
22821 Lake Forest Dr.
Suite 114
Lake Forest, CA 92630
(714) 770-9616

Eugene D. Finkle, MD
P.O. Box 309
Laytonville, CA 95454
(707) 984-6151

H. Richard Casdorph, MD
1703 Termino Ave., #201
Long Beach, CA 90804
(310) 597-8716

Robert F. Cathcart III, MD
127 Second St., #4
Los Altos, CA 94022
(415) 949-2822

Claude Marquette, MD
5050 El Camino Real, #110
Los Altos, CA 94022
(415) 964-6700

Laszlo Belenyessy, MD
12732 Washington Blvd, #D
Los Angeles, CA 90066
(213) 822-4614

M. Jahangiri, MD
2156 S. Santa Fe
Los Angeles, CA 90058
(213) 587-3218

Lon B. Work, MD
841 Foam St., #D
Monterrey, CA 93940
(408) 655-0215

Julian Whitaker, MD
4400 MacArthur Blvd.
Suite 630
Newport Bch., CA 92660
(714) 851-1550

David C. Freeman, MD
11311 Camarillo St., #103
N. Hollywood, CA 91602
(818) 985-1103

A. Hal Thatcher, MD
2552 Cornwall St.
Oceanside, CA 92054
RETIRED

Mohamed Moharram, MD
300 W. 5th Street, #B
Oxnard, CA 93030
(805) 483-2355

David H. Tang, MD
74133 El Paseo, #6
Palm Desert, CA 92260
(619) 341-2113

Sean Degnan, MD
2825 Tahqultz McCallum
Suite 200
Palm Springs, CA 92262
(619) 320-4292

John B. Park, MD
131 East Mill Avenue
Porterville, CA 93527
(209) 781-6224

Charles Farinella, MD
69-730 Hwy. 111
Suite 106A
Rancho Mirage, CA 92270
(619) 324-0734

Bessie J. Tillman, MD
2054 Market St.
Redding, CA 96001
(916) 246-3022

Ilona Abraham, MD
19231 Victoria Blvd.
Reseda, CA 91335
(818) 345-8721

J.E. Dugas, MD
3400 Cottage Way, #206
Sacramento, CA 95825
RETIRED

Michael Kwiker, DO
3301 Alta Arden, Ste. 3
Sacramento, CA 95825
(916) 489-4400

William Doell, DO
3301 Calle Negocio
San Clemente, CA 92672
NO REFERRALS

Lawrence Taylor, MD
3330 Third Ave., #402
San Diego, CA 92103
(619) 296-2952

Richard A. Kunin, MD
2698 Pacific Avenue
San Francisco, CA 94115
(415) 346-2500

Russel A. Lemesh, MD
595 Buckingham Way
#320
San Francisco, CA 94132
(415) 731-5907

Paul Lynn, MD
345 W. Portal Ave.
San Francisco, CA 94127
(415) 566-1000

Gary S. Ross, MD
500 Sutter, #300
San Francisco, CA 94102
(415) 398-0555

Steven H. Gee, MD
595 Estudillo St.
San Leandro, CA 94577
(510) 483-5881

Ross B. Gordon, MD
4144 Redwood Highway
San Rafael, CA 94903
(415) 499-9377

William C. Kubitschek, DO
1194 Calle Maria
San Marcos, CA 92069
(619) 744-6991

Ronald Wempen, MD
3620 S. Bristol Street, #306
Santa Ana, CA 92704
(714) 546-4325

H.J. Hoegerman, MD
101 W. Arrellaga, Ste. D
Santa Barbara, CA 93101
(805) 963-1824

Mohamed Moharram, MD
101 W. Arrellaga, Ste. B
Santa Barbara, CA 93101
(805) 965-5229

Donald E. Reiner, MD
1414-D. South Miller
Santa Maria, CA 93454
(805) 925-0961

Michael Rosenbaum, MD
2730 Wilshire Blvd., #110
Santa Monica, CA 90403
(310) 453-4424

Murray Susser, MD
2730 Wilshire Blvd., #110
Santa Monica, CA 90403
(310) 453-4424

Terri Su, MD
1038 4th Street, #3
Santa Rosa, CA 95404
(707) 571-7560

Allen Green, MD
909 Electric Ave., #212
Seal Beach, CA 90740
(310) 493-4526

Rosa M. Ami Belli, MD
13481 Cheltenham Dr.
Sherman Oaks, CA 91423
NO REFERRALS

JoAnn Hoffer, MD
12559 Hwy. 101 (mini mart)
Smith River, CA 95567
(707) 487-3405

James D. Schuler, MD
12599 Hwy. 101 (mini mart)
Smith River, CA 95567
(707) 487-3405

William J. Goldwag, MD
7499 Cerritos Ave.
Stanton, CA 90680
(714) 827-5180

Charles E. Law Jr., MD
3959 Laurel Canyon Blvd.
Suite 1
Studio City, CA 91604
(818) 761-1661

Anita Millen, MD
1010 Crenshaw Blvd.
Suite 170
Torrance, CA 90501
(310) 320-1132

Frank Mosler, MD
14428 Gilmore St.
Van Nuys, CA 91401
(818) 785-7425

Alan Shifman Charles, MD
1414 Maria Lane
Walnut Creek, CA 94596
(510) 937-3331

Peter H.C. Mutke, MD
1808 San Miguel Drive
Walnut Creek, CA 94596
(510) 933-2405

Colorado

James R. Fish, MD
3030 N. Hancock
Colo. Springs, CO 80907
(719) 471-2273

George Juetersonke, DO
5455 N. Union, #200
Colo. Springs, CO 80918
(719) 528-1960

John H. Altshuler, MD
Greenwood Exec. Park, Bld. 10
7485 E. Peakview Avenue
Englewood, CO 80111
(303) 740-7771

William L. Reed, MD
591- 25 Road, Ste. A-4
Grand Junction, CO 81505
(303) 241-3631

Connecticut

Jerrold N. Finnie, MD
333 Kennedy Dr., #204
Torrington, CT 06790
(203) 489-8977

District of Columbia

Paul Beals, MD
2639 Connecticut Ave., NW
Suite 100
Washington, D.C. 20037
(202) 332-0370

George H. Mitchell, MD
2639 Connecticut Ave., NW
Suite C-100
Washington, D.C. 20008
(202) 265-4111

Florida

Richard Worsham, MD
303 - 1st Street
Atlantic Beach, FL 32233
NO REFERRALS

Leonard Haimes, MD
7300 N. Federal Hwy., Ste 107
Boca Raton, FL 33487
(407) 994-3868

Narinder Singh Parhar, MD
7840 Glades Road, #220
Boca Raton, FL 33434
(407) 479-3200

Eteri Meinikov, MD
116 Manatee Ave. East
Brandenton, FL 34208
(813) 748-7943

Bruce Dooley, MD
1493 S.E. 17th Street
Fort Lauderdale, FL 33316
(305) 527-9355

Gary L. Pynckel, DO
3940 Metro Parkway, #115
Fort Meyers, FL 33916
(813) 278-3377

Herbert Pardell, DO
7061 Taft Street
Hollywood, FL 33020
(305) 989-5558

Carlos F. Gonzalez, MD
7991 S. Suncoast Blvd.
Homosassa, FL 32646
(904) 382-8282

Neil Ahner, MD
1080 E. Indiantown Rd.
Jupiter, FL 33477
(407) 744-0077

Harold Robinson, MD
4406 S. Florida Ave.
Suite 27
Lakeland, FL 33803
(813) 646-5088

Herbert S. Slavin, MD
7200 W. Commercial Blvd.
Suite 210
Lauderhill, FL 33319
(305) 748-4991

Joya Lynn Schoen, MD
341 N. Maitland Ave.
Suite 200
Maitland, FL 32751
(407) 644-2729

Joseph G. Godorov, DO
9055 S.W. 87th Ave.
Suite 307
Miami, Fl 33176
(305) 595-0671

Bernard J. Letourneau, DO
6475 SW 40th Street
Miami, FL 33155
(305) 666-9933

Narinder Singh Parhar, MD
1333 S. State Road 7
N. Lauderdale, FL 33068
(305) 978-6604

Martin Dayton, DO
18600 Collins Avenue
N. Miami Bch., FL 33160
(305) 931-8484

George Graves, DO
3501 N.E. Tenth St.
Ocala, FL 32670
(904) 236-2525

Travis L. Herring, MD
106 West Fern Drive
Orange City, FL 32763
(904) 775-0525

Neil Ahner, MD
1200 Malabar Road
Palm Bay, FL 32907
(407) 729-8581

Ward Dean, MD
P.O. Box 11097
Pensacola, FL 32524
NO REFERRALS

Dan C. Roehm, MD
3400 Park Central Blvd. N.
Suite 3450
Pompano Bch., FL 33064
(305) 977-3700

James Parsons, MD
707 Mullet Dr., Ste. 110
Port Cavernal, FL 32920
(407) 784-2102

Joseph Ossorlo, MD
3900 Clark Rd., #H-5
Sarasota, FL 34277
(813) 921-6338

Ray Wunderlich Jr., MD
666 - 6th Street South
St. Petersburg, FL 33701
(813) 822-3612

Leon L. Shore, DO
10111 W. Oakland Park Blvd.
Sunrise, FL 33351
(305) 741-1533

Donald J. Carrow, MD
3902 Henderson Blvd.
Suite. 206
Tampa, FL 33629
(813) 832-3220

Eugene H. Lee, MD
1804 W. Kennedy Blvd., #A
Tampa, FL 33606
(813) 251-3089

Thomas McNaughton, MD
540 S. Nokomis Ave.
Venice, FL 34285
(813) 484-2167

Alfred S. Massam, MD
528 West Main Street
Wauchula, FL 33873
(813) 773-6668

James M. Parsons, MD
Great Western Bank Bldg.
#303, 2699 Lee Road
Winter Park, FL 32789
(407) 628-3399

Robert Rogers, MD
1865 N. Semoran Blvd.
Suite 204
Winter Park, FL 32792
(407) 679-2811

Georgia

David Epstein, DO
427 Moreland Ave., #100
Atlanta, GA 30307
(404) 525-7333

Milton Fried, MD
4426 Tilly Mill Road
Atlanta, GA 30360
(404) 451-4857

Bernard Mlaver, MD
4480 N. Shallowford Rd.
Atlanta, GA 30338
(404) 395-1600

Oliver L Gunter, MD
24 N. Ellis St.
Camilla, GA 31730
(912) 336-7343

Naima ABD Elghany, MD
3455H. N. Druid Hills Rd.
Decatur, GA 30033
(404) 639-3385

Stephen Edelson, MD
6920 Jimmy Carter Blvd.
Norcross, GA 30071
(404) 729-8359

Terril J. Schneider, MD
205 Dental Drive, Ste. 19
Warner Robbins, GA 31088
(912) 929-1027

Hawaii

Clifton Arrington, MD
P.O. Box 649
Kealakekua, HI 96750
(808) 322-9400

Idaho

Charles T. McGee, MD
1717 Lincoln Way
Suite 108
Coeur d'Alene, ID 83814
(208) 664-1478

John O. Boxall, MD
824 - 17th Ave. South
Nampa, ID 83651
(208) 466-3517

Stephen Thornburgh, DO
824- 17th Ave. South
Nampa, ID 83651
(208) 466-3517

K. Peter McCallum, MD
2500 Selle Road
Sandpoint, ID 83864
(208) 263-5456

Illinois

Terrill K. Haws, DO
121 S. Wilke Road
Suite 111
Arlington Hts., IL 60005
(708) 577-9451

William J. Mauer, DO
3401 N. Kennicott Ave.
Arlington Hts., IL 60004
(800) 255-7030

Thomas Hesselink, MD
888 S. Edgelawn Dr.
Suite 1735
Aurora, IL 60506
(708) 844-0011

M. Paul Dommers, MD
554 S. Main Street
Belvidere, IL 61008
(815) 544-3112

Razvan Rentea, MD
3354 N. Paulina
Chicago, IL 80657
(312) 549-0101

Guillermo Justiniano, MD
1430 Parish Court
Downers Grove, IL 60515
(708) 964-8083

Richard E. Hrdlicka, MD
302 Randall Rd., #206
Geneva, IL 60134
(708) 232-1900

Robert S. Waters, MD
739 Roosevelt Road
Glen Ellyn, IL 60137
(708) 790-8100

Frederick Weiss, MD
3207 W. 184th Street
Homewood, IL 60430
NO REFERRALS

Stephen K. Elsasser, DO
205 S. Englewood
Metamora, IL 61548
(309) 367-2321

Terry W. Love, DO
2610 - 41st Street
Moline, IL 61252
(309) 764-2900

Paul J. Dunn, MD
715 Lake St.
Oak Park, IL 60301
(708) 383-3800

Terry W. Love, DO
645 W. Main
Ottawa, IL 61350
(815) 434-1977

John R. Tambone, MD
102 E. South St.
Woodstock, IL 60098
(815) 338-2345

Peter Senatore, DO
1911 - 27th St.
Zion, IL 60099
(708) 872-8722

Indiana

George Wolverton, MD
647 Eastern Blvd.
Clarksville, IN 47130
(812) 282-4309

Harold T. Sparks, DO
3001 Washington Ave.
Evansville, IN 47714
(812) 479-8228

Cal Streeter, DO
9635 Saric Court
Highland, IN 46322
(219) 924-2410

David A. Darbro, MD
2124 E. Hanna Ave.
Indianaoplis, IN 46227
(317) 787-7221

Norman E. Whitney, DO
P.O. Box 173
Morreseville, IN 46158
(317) 831-3352

David E. Turfler, DO
336 W. Navarre St.
South Bend, IN 46616
(219) 233-3840

Myrna D. Trowbridge, DO
850-C Marsh St.
Valparaiso, IN 46383
(219) 462-3377

Iowa

Beverly Rosenfeld, DO
7177 Hickman Rd., #10
Des Moines, IA 50322
(515) 276-0061

Horst G. Blume, MD
700 Jennings St.
Sioux City, IA 51105
(712) 252-4386

Kansas

Stevens B. Acker, MD
310 West Central, #D
P.O. Box 483
Andover, KS 67002
(316) 733-4494

Terry Hunsberger, DO
602 N. 3rd - P.O. Box 679
Garden City, KS 67846
(316) 275-7128

Roy N. Neil, MD
105 West 13th
Hays, KS 67601
(913) 628-8341

John Gamble, Jr., DO
1509 Quindaro
Kansas City, KS 66104
(913) 321-1140

Kentucky

Edward K. Atkinson, MD
P.O. Box 3148
Berea, KY 40403
NO REFERRALS

John C. Tapp, MD
414 Old Morgantown Rd.
Bowling Green, KY 42101
(502) 781-1483

Kirk Morgan, MD
9105 U.S. Hwy. 42
Louisville, KY 40059
(502) 228-0158

Walt Stoll, MD
6801 Danville Road
Nicholasville, KY 40356
(606) 233-4273

Stephen S. Kiteck, MD
1301 Pumphouse Rd.
Somerset, KY 42501
(606) 678-5137

Louisisana

Steve Kuplesky, MD
5618 Bayridge
Baton Rouge, LA 70817
NO REFERRALS

Saroj T. Tampira, MD
812 E. Judge Perez
Chalmette, LA 70043
(504) 277-8991

Roy M. Montalbano, MD
4408 Highway 22
Mandeville, LA 70448
(504) 626-1985

Phillip Mitchell, MD
407 Blenville St.
Natchitoches, LA 71457
(800) 562-6574

Joseph R. Whitaker, MD
P.O. Box 458
Newellton, LA 71357
(318) 467-5131

Adonis J. Domingue, MD
602 N. Lewis, #600
New Iberia, LA 70580
(318) 365-2196

James P. Carter, MD
1430 Tulane Avenue
New Orleans, LA 70112
(504) 588-5136

R. Denman Crow, MD
1545 Line Ave., Ste. 222
Shreveport, LA 71101
(318) 221-1569

Maine

Joseph Cyr, MD
62 Main Street
Van Buren, ME 04785
(207) 868-5273

Maryland

Paul V. Beals, MD
9101 Cherry Lane Park
Suite 205
Laurel, MD 20708
(301) 490-9911

Alan R. Gaby, MD
31 Walker Ave.
Pikesville, MD 21208
(410) 486-5656

Harold Goodman, DO
121 Congressional Ln.
Suite 208
Rockville, MD 20852
(301) 881-5229

Massachusetts

Michael Janson, MD
275 Mill Way
P.O. Box 732
Barnstable, MA 02630
(508) 362-4343

Michael Janson, MD
2557 Massachusetts Ave.
Cambridge, MA 02140
(617) 661-6225

Richard Cohen, MD
51 Mill St., #1
Hanover, MA 02339
(617) 829-9281

Svetlana Kaufman, MD
24 Merrimack St., #323
Lowell, MA 01852
(508) 453-5181

Carol Englender, MD
1340 Centre St.
Newton, MA 02159
(617) 965-7770

N. Thomas La Cava, MD
360 W. Boylston St.
Suite. 107
West Boylston, MA 01583
(508) 854-1380

Ross S. McConnell, MD
732 Main St.
Williamstown, MA 01267
(413) 663-3701

Michigan

Leo Modzinski, DO, MD
100 W. State St.
Atlanta, MI 49709
(517) 785-4254

Doyle B. Hill, DO
2520 N. Euclid Avenue
Bay City, MI 48706
(517) 686-5200

Paul A. Parente, DO
30275 Thirteen Mile Rd.
Farmington Hills, MI 48018
(313) 626-9690

Albert J. Scarchilli, DO
30274 Thirteen Mile Rd.
Farmington Hills, 48018
(313) 626-9690

William M. Bernard, DO
1044 Gilbert St.
Flint, MI 48532
(313) 733-3140

Kenneth Ganapini, DO
1044 Gilbert St.
Flint, MI 48532
(313) 733-3140

E. Duane Powers, DO
P.O. Box 170
Grand Haven, MI 49417
RETIRED

Grant Born, DO
2687 - 44th St. SE
Grand Rapis, MI 49512
(616) 455-3550

Marvin D. Penwell, DO
319 S. Bridge St.
Linden, MI 48451
(313) 735-7809

Vahagn Agbabian, DO
28 N. Saginaw St.
Suite 1105
Pontiac, MI 48058
(313) 334-2424

Richard E. Tapert, DO
23550 Harper
St. Clair Shores, MI 48080
(313) 779-5700

Seldon Nelson, DO
4386 N. Meridian Rd.
Williamston, MI 48895
(517) 349-2458

Minnesota

Michael Dole, MD
10700 Old County Rd. 15
Suite 350
Minneapolis, MN 55441
(612) 593-9458

Jean R. Eckerly, MD
10700 Old County Rd. 15
Suite 350
Minneapolis, MN 55441
(612) 593-9458

Keith J. Carlson, MD
210 Highland Ct.
Tyler, MN 56178
(507) 247-5921

F.J. Durand, MD
3119 Groveland School Rd.
Wayzata, MN 55391
NO REFERRALS

M.S.C. Durand, MD
3119 Groveland School Rd.
Wayzata, MN 55391
NO REFERRALS

Mississippi

Pravinchandra Patel, MD
P.O. Drawer DD
Coldwater, MS 38618
(601) 622-7011

James H. Sams, MD
1120 Lehmburg Rd.
Columbus, MS 39702
(601) 327-8701

James H. Waddell, MD
1520 Government St.
Ocean Springs, MS 39564
(601) 875-5505

Robert Hollingsworth, MD
Drawer 87, 901 Forrest St.
Shelby, MS 38774
(601) 398-5106

Missouri

John T. Schwent, DO
1400 Truman Blvd.
Festus, MO 63028
(314) 937-8688

Tipu Sultan, MD
11585 W. Florissant
Florissant, MO 63033
(314) 921-7100

Lawrence Dorman, DO
9120 E. 35th St.
Independence, MO 64052
(816) 358-2712

James E. Swann, DO
2116 Sterling
Independence, MO 64052
(816) 833-3366

Edward W. McDonagh, DO
2800-A Kendallwood Pkwy.
Kansas City, MO 64119
(816) 453-5940

James Rowland, DO
8133 Wornall Rd.
Kansas City, MO 64114
(816) 361-4077

Charles J. Rudolph, DO, PhD
2800-A Kendallwood Pkwy.
Kansas City, MO 64119
(816) 453-5940

William C. Sunderwirth, DO
2828 N. National
Springfield, MO 65803
(417) 869-6260

Harvey Walker, Jr., MD, PhD
138 N. Meramec Avenue
St. Louis, MO 63105
(314) 721-7227

William C. Sunderwirth, DO
307 South St.
Stockton, MO 65785
(417) 276-3221

Ronald H. Scott, DO
131 Meredith Lane
Sullivan, MO 63080
(314) 468-4932

Clinton C. Hayes, DO
100 W. Main
Union, MO 63084
(314) 583-8911

Nebraska

Eugene C. Oliveto, MD
8031 W. Center Rd., #208
Omaha, NE 68124
(402) 392-0233

Otis W. Miller, MD
408 S. 14th Street
Ord, NE 68862
(308) 728-3251

Nevada

W. Douglas Brodie, MD
848 Tanager
Incline Village, NV 89450
(702) 832-7001

Ji-Zhou (Joseph) Kang, MD
5613 S. Eastern
Las Vegas, NV 89119
(702) 798-2992

Paul McGuff, MD
3930 Swenson, #903
Las Vegas, NV 89106
RETIRED

Robert D. Milne, MD
501 S. Rancho, Ste. 446
Las Vegas, NV 89106
(702) 385-1999

Terry Pfau, DO
501 S. Rancho Dr., Ste.446
Las Vegas, NV 89106
(702) 385-1999

Robert Vance, DO
801 S. Rancho Dr., Ste. F2
Las Vegas, NV 89106
(702) 385-7771

David A. Edwards, MD
6490 S. McCarran Bl.
Suite A7
Reno, NV 89509
(702) 827-1444

Michael L. Gerber, MD
3670 Grant Drive
Reno, NV 89509
(702) 826-1900

Donald E. Soll, MD
708 N. Center St.
Reno, NV 89501
(702) 786-7101

Yiwen Y. Tang, MD
380 Brinkby
Reno, NV 89509
(702) 826-9500

New Jersey

Majid Ali, MD
320 Belleville Ave.
Bloomfield, NJ 08003

Allan Magaziner, DO
1907 Greentree Rd.
Cherry Hill, NJ 08003
(609) 424-8222

Majid Ali, MD
95 E. Main St.
Denville NJ 07834
(201) 586-4111

C.Y. Lee, MD
952 Amboy Avenue
Edison, NJ 08837
(908) 738-9220

Ralph Lev, MD
952 Amboy Avenue
Edison, NJ 08837
(908) 738-9220

Richard B. Menashe, DO
15 S. Main Street
Edison, NJ 08837
(908) 906-8866

Gennaro Locurcio, MD
610 - 3rd Avenue
Elizabeth, NJ 07202
(908) 351-1333

Charles Harris, MD
1 Ortley Plaza
Ortley Beach, NJ 08751
(908) 793-6464

Constance Alfano, MD
74 Oak Street
Ridgewood, NJ 07450
(201) 444-4622

Eric Braverman, MD
100-102 Tamarck Circle
Skillman, NJ 08558
(609) 921-1842

Faina Munits, MD
51 Pleasant Valley Way
West Orange, NJ 07052
(201) 736-3743

New Mexico

Ralph J. Luciani, DO
2301 San Pedro NE.
Suite G
Albuquerque, NM 87110
(505) 888-5995

Gerald Parker, DO
6208 Montgomery Blvd. NE
Suite D
Albuquerque, NM 87109
(505) 884-3506

John T. Taylor, DO
6208 Montgomery Blvd. NE
Suite D
Albuquerque, NM 87109
(505) 884-3506

Annette Stoesser, MD
112 S. Kentucky
Roswell, NM 88201
(505) 623-2444

New York

Richard Izquierdo, MD
1070 Southern Blvd.
Lower Level
Bronx, NY 10459
(212) 589-4541

Gennaro Locurcio, MD
2386 Ocean Parkway
Brooklyn, NY 11223
(718) 336-2291

Tsillia Sorina, MD
2026 Ocean Avenue
Brooklyn, NY 11230
(718) 375-2600

Michael Teplitsky, MD
415 Oceanview Avenue
Brooklyn, NY 11235
(718) 769-0997

Pavel Yutsis, MD
1309 W. 7th St.
Brooklyn, NY 11204
(718) 259-2122

Christopher Calapal, DO
1900 Hempstead Tnpke.
East Meadow, NY 11554
(516) 794-0404

Reino Hill, MD
230 West Main St.
Falconer, NY 14733
(716) 665-3505

Mary F. Di Rico, MD
1 Kingspoint Rd.
Greak Neck, NY 11024
NO REFERRALS

Serafina Corsello, MD
175 E. Main St.
Huntington, NY 11743
(516) 271-0222

Mitchell Kurk, MD
310 Broadway
Lawrence, NY 11559
(516) 239-5540

Bob Snider, MD
HC 61, Box 43D
Massena, NY 13662
(315) 764-7328

Robert C. Atkins, MD
152 E. 55th St.
New York, NY 10022
(212) 758-2110

Serafina Corsello, MD
200 W. 57th St., #1202
New York, NY 10019
(212) 399-0222

Ronald Hoffman, MD
40 E. 30th St.
New York, NY 10016
(212) 779-1744

Warren M. Levin, MD
444 Park Ave. So./30th St.
New York, NY 10016
(212) 696-1900

Paul Cutler, MD
652 Elmwood Avenue
Niagra Falls, NY 14301
(716) 284-5140

Neil L. Block, MD
14 Prel Plaza
Orangeburg, NY 10962
(914) 359-3300

Driss Hassam, MD
50 Court Street
Plattsburgh, NY 12901
(518) 561-2023

Kenneth A. Bock, MD
108 Montgomery St.
Rhinebeck, NY 12572
(914) 876-7082

Michael B. Schachter, MD
Two Executive Blvd., #202
Suffern, NY 10901
(914) 368-4700

Rodolfo T. Sy, MD
1845 - 6th Avenue
Watervilet, NY 12189
(518) 273-1325

Savely Yurkovsky, MD
309 Madison St.
Westbury, NY 11590
(516) 333-2929

North Carolina

Keith E. Johnson, MD
188 Quewhiffle
Aberdeen, NC 28315
(919) 281-5122

John L. Laird, MD
Rt. 1- Box 7
Leicester, NC 28748
(704) 683-3101

John L. Laird, MD
Plaza 21 North
Statesville, NC 28677
(704) 876-1617

North Dakota

Richard H. Leigh, MD
2314 Library Circle
Grand Forks, ND 58201
(701) 775-5527

Brian E. Briggs, MD
718 - 6th Street S.W.
Minot, ND 58701
(701) 838-6011

Ohio

Francis J. Walckman, MD
544B. White Pond Drive
Akron, OH 44320
(216) 867-3767

L. Terry Chappell, MD
122 Thurman St.
Bluffton, OH 45817
(419) 358-4627

Jack E. Slingluff, DO
5850 Fulton Rd., NW
Canton, OH 44718
(216) 494-8641

Ted Cole, DO
9678 Cincinnati-Columbus Rd.
Cincinnati, OH 45241
(513) 779-0300

John M. Baron, DO
4807 Rockside, Ste. 100
Cleveland, OH 44131
(216) 642-0082

James P. Frackelton, MD
24700 Center Ridge Rd.
Cleveland, OH 44145
(216) 835-0104

Derrick Lonsdale, MD
24700 Center Ridge Rd.
Cleveland, OH 44145
(216) 835-0104

Douglas Weeks, MD
24700 Center Ridge Rd.
Cleveland, OH 44145
(216) 835-0104

Robert R. Hershner, DO
1571 E. Livingston Ave.
Columbus, OH 43255
(614) 253-8733

William D. Mitchell, DO
3520 Snouffer Rd.
Columbus, OH 43235
(614) 761-0555

David G. Goldberg, DO
100 Forest Park Dr.
Dayton, OH 45405
(513) 277-1722

Richard Sielski, MD
3484 Cincinnati-Zainsville Rd.
Lancaster, OH 43130
(614) 653-0017

Don K. Snyder, MD
Route 2 - Box 1271
Paulding, OH 45879
(419) 399-2045

James Ventresco, Jr., DO
3848 Tippecanoe Rd.
Youngstown, OH 44511
(216) 792-2349

Oklahoma

Leon Anderson, DO
121 Second St.
Jenk, OK 74037
(918) 299-5039

Charles H. Farr, MD, PhD
8524 S. Western, Ste. 107
Oklahoma City, OK 73139
(405) 632-8868

Charles D. Taylor, MD
3715 N. Classen Blvd.
Oklahoma City, OK 73118
(405) 525-7751

Oregon

Ronald L. Peters, MD
1607 Siskiyou Blvd.
Ashland, OR 97520
(503) 482-7007

John Gambee, MD
66 Club Road, Ste. 140
Eugene, OR 97401
(503) 686-2536

James Fitzsimmons, Jr., MD
591 Hidden Valley Rd.
Grants Pass, OR 97527
(503) 474-2166

Terence Howe Young, MD
1205 Wallace Rd, NW
Salem, OR 97304
(503) 371-1558

Pennsylvania

Robert H. Schmidt, DO
1227 Liberty Plaza Bldg.
Suite. 303
Allentown, PA 18102
(215) 437-1959

D. Erik Von Kiel, DO
Libert Square Med. Center
Suite 200
Allentown, PA 18104
(215) 776-7639

Francis J. Cinelli, DO
153 N. 11th St.
Bangor, PA 18013
(215) 588-4502

Bill Illingworth, DO
120 West John St.
Bedford, PA 15522
(814) 623-8414

Sally Ann Rex, DO
1343 Easton Ave.
Bethlehem, PA 18018
(215) 868-0900

Dennis L. Gilbert, DO
50 N. Market St.
Elizabethtown, PA 17022
(717) 367-1345

Harold H. Byer, MD, PhD
5045 Swamp Rd., #A-101
Fountainville, PA 18923
(215) 348-0443

Ralph A. Miranda, MD
RD #12 - Box 108
Greensburg, PA 15801
(412) 838-7632

Arthur L. Koch, DO
57 West Juniper St.
Hazelton. PA 18201
(717) 455-4747

Chandrika Sinha, MD
1177 S. Sixth St.
Indiana, PA 15701
(412) 349-1414

D. Erik Von Keil, DO
7386 Alburtis Rd., Ste.101
Macungle, PA 18062
(215) 967-5503

Conrad G. Maulfair, Jr., DO
Box 71 - Main St.
Mertztown, PA 19539
(215) 682-2104

Mamduh El-Attrache, MD
20 E. Main St.
Mt. Pleasant, PA 15666
(412) 547-3576

Mamduh El-Attrache, MD
215 Crooked Run Road
North Versailles, PA 15137
(412) 673-3900

Robert J. Peterson, DO
64 Magnolia Drive
Newtown, PA 18940
NO REFERRALS

Frederick Burton, MD
69 W. Schoolhouse Lane
Philadelphia, PA 19144
(215) 844-4660

Jose Castillo, MD
228 South 22nd St.
Philadelphia, PA 19103
(215) 567-5845

Mura Galperin, MD
824 Hendrix St.
Philadelphia, PA 19116
(215) 677-2337

P. Jayalakshmi, MD
6366 Sherwood Road
Philadelphia, PA 19151
(215) 473-4226

K.R. Sampathachar, MD
6366 Sherwood Road
Philadelphia, PA 19151
(215) 473-4226

Lance Wright, MD
3901 Market Street
Philadelphia, PA 19104
(215) 387-1200

Harold Buttram, MD
5724 Clymer Road
Quakertown, PA 18951
(215) 536-1890

Paul Peirsel, MD
RD 4 Box 257A
Somerset, PA 15541
(814) 443-2521

South Carolina

Theodore C. Rozema, MD
2228 Airport Road
Columbia, SC 29205
(803) 796-1702

Theodore C. Rozema, MD
1000 E. Rutherford Rd.
Landrum, SC 29356
(803) 457-4141

Tennessee

S. Marshall Fram, MD
135 Weatheridge Dr.
Jackson, TN 38305
RETIRED

Donald Thompson, MD
P.O. Box 2088
Morristown, TN 37816
(615) 581-8367

Stephen L. Reisman, MD
417 East Iris Dr.
Nashville, TN 37204
(615) 383-9030

Texas

Herbert Carr, DO
P.O. Box 1179
Alamo, TX 78516
(512) 787-6668

William Irby Fox, MD
1227 N. Mockingbird Ln.
Abilene, TX 79603
(915) 672-7863

Gerald Parker, DO
4714 S. Western
Amarillo, TX 79109
(806) 355-8263

John T. Taylor, DO
4714 S. Western
Amarillo, TX 79109
(806) 355-8263

Vladimir Rizov, MD
8311 Shoal Creek Blvd.
Austin, TX 78758
(512) 451-8149

Brij Myer, MD
4222 Trinity Mills Rd.
Suite 222
Dallas, TX 75287
(214) 248-2488

Michael G. Samuels, DO
7616 LBJ Freeway, #230
Dallas, TX 75251
(214) 991-3977

J. Robert Winslow, DO
2745 Valwood Pkwy.
Dallas, TX 75234
(214) 241-4614

J. Robert Winslow, DO
2815 Valley View Lane
Suite 111
Dallas, TX 75234
(214) 243-7711

Edward J. Etti, MD
3500 North Piedras
P.O. Box 31397
El Paso, TX 79931
(915) 566-9361

Francisco Soto, MD
424 Executive Center Blvd.
Suite 100
El Paso, TX 79902
(915) 534-0272

Robert Battle, MD
9910 Long Point
Houston, TX 77055
(713) 932-0552

Jerome L. Borochoff, MD
8830 Long Point
Suite 504
Houston, TX 77055
(713) 461-7517

Luis E. Guerrero, MD
2055 S. Gessner, Suite 150
Houston, TX 77063
(713) 789-0133

Carlos E. Nossa, MD
3800 Tanglewilde
Suite 1007
Houston, TX 77063
NO REFERRALS

John P. Trowbridge, MD
9816 Memorial Blvd.
Suite 205
Humble, TX 77338
(713) 540-2329

John L. Sessions, DO
1609 South Margaret
Kirbyville, TX 75956
(409) 423-2166

Ronald M. Davis, MD
10414 W. Main St.
La Porte, TX 77571
(713) 470-2930

Ruben Berlanga, MD
649-B Dogwood
Laredo, TX 78041
NO REFERRALS

Ricardo Tan, MD
423 S. Palm
Pecos, TX 79772
(915) 445-9090

Linda Martin, DO
1524 Independence, #C
Plano, TX 75075
(214) 985-1377

Jim P. Archer, DO
8434 Fredericksburg Rd.
San Antonio, TX 78229
(512) 615-8445

Ron Stogryn, MD
7334 Blanco Rd., #100
San Antonio, TX 78216
(512) 366-3637

Thomas R. Humphrey, MD
2400 Rushing
Wichita Falls, TX 76308
(817) 766-4329

Utah

Dennis Harper, DO
1675 N. Freedom Blvd.
Suite 11E
Provo, UT 84604
(801) 373-8500

D. Remington, MD
1675 N. Freedom Blvd.
Suite 11E
Provo, UT 84604
(801) 373-8500

Virginia

Sohini Patel, MD
7023 Little River Tnpk.
Suite 207
Annandale, VA 22003
(703) 941-3606

Harold Huffman, MD
P.O. Box 197
Hinton, VA 22831
(703) 867-5242

Peter C. Gent, DO
11900 Hull Street
Midlothian, VA 23112
(804) 744-3551

Vincent Speckhart, MD
902 Graydon Ave.
Norfolk, VA 23507
(804) 622-0014

Elmer M. Cranton, MD
Ripshin Road - Box 44
Trout Dale, VA 24378
(703) 677-3631

Washington

David Buscher, MD
1370 - 116th N.E.
Suite 102
Bellevue, WA 98004-3825
(206) 453-0288

Maurice Stephens, MD
5011 133rd. Pl. SE
Bellevue, WA 98006
NO REFERRALS

Robert Kimmel, MD
4204 Meridian, Ste. 104
Bellingham, WA 98226
(206) 734-3250

James P. DeSanits, DO
8116 Palm St.
Fairchild AFB, WA 99011
NO REFERRALS

Jonathan Wright, MD
24030 - 132nd S.E.
Kent, WA 98042
(206) 631-8920

Jonathan Collin, MD
12911 - 128th St., N.E.
Suite F-100
Kirkland, WA 98034
(206) 820-0547

Jonathan Collin, MD
911 Tyler St.
Port Townsend, WA 98368
(206) 385-4555

Michael G. Vasselago, MD
217 North 125th
Seattle, WA 98133
(206) 367-0760

Burton B. Hart, DO
E. 12104 Main
Spokane, WA 99206
(509) 927-9922

Richard P. Huemer, MD
406 S.E. 131st Ave.
Building.C-303
Vancouver, WA 98684
(206) 253-4445

Murray L. Black, DO
609 S. 48th Ave.
Yakima, WA 98908
(509) 966-1780

Elmer M. Cranton, MD
15246 Leona Dr., S.E.
Yelm, WA 98597
(206) 894-3548

West Virginia

Prudencio Corro, MD
251 Stanaford Rd.
Beckley, WV 25801
(304) 252-0775

Michael Kostenko, DO
114 E. Main St.
Beckley, WV 25801
(304) 253-0591

Steve M. Zekan, MD
1208 Kanawha Blvd., E
Charleston, WV 25301
(304) 343-7559

Wisconsin

Eleazar M. Kadile, MD
1538 Bellevue St.
Green Bay, WI 54311
(414) 468-9442

Rathna Alwa, MD
717 Geneva St.
Lake Geneva, WI 53147
(414) 248-1430

William J. Faber, DO
6529 W. Fond du Lac Ave.
Milwaukee, WI 53218
(414) 464-7680

Thomas Hesselink, MD
10520 W. Blue Mound Rd.
Suite 202
Milwaukee, WI 53226
(414) 259-1350

Robert R. Stocker, DO
2505 N. Mayfair Rd.
Milwaukee, WI 53226
RETIRED

Jerry N. Yee, DO
2505 N. Mayfair Rd.
Milwaukee, WI 53226
(414) 258-6282

Robert S. Waters, MD
Race & Vine Streets
Box 357
Wisconsin Dells, WI 53965
(608) 254-7178

APPENDIX II

Medicare Reimbursement Sample Letter

Example of a letter seeking reimbursement by Medicare for receiving chelation therapy written before or after treatment and using the response as the basis of your claim made after treatment.

Heading and date

Addressed to the Social Security Administration
Office in your city, the Medicare Claims Division

Dear Administrator of Medicare Claims:

I request your permission and assurance of reimbursement under Parts A and B of my Medicare Health Insurance Coverage Policy Number ________ allowing me to receive chelation therapy with ethylene diamine tetraacetic acid for atherosclerosis and allied health problems.

According to a policy decision by the Health Care Financing Administration, U.S. Dept. of Health, Education, and Welfare, published in the Medicare Carriers Manual, Part B, Part 3, transmittal No. 799—April 1980, distributed to all Medicare Regional Offices on May 15, 1980, you were informed that on that date the following change became effective:

> SECTION 2050. 5—*Drugs and Biologicals*, and the *Coverage Issues Appendix*, have been revised to add new guidance for determining drug coverage, intended to facilitate contractors' paying for drugs sooner after FDA approval than is now done. As before, a drug or biological approved by FDA is considered safe and effective, and may be paid for if other applicable coverage requirements are met. The new policy permits contractors to pay for drugs after obtaining satisfactory evidence of FDA approval in questionable cases such as those of new drugs only recently approved. In addition, coverage of FDA-approved drugs is extended to any use except those expressly disapproved by the FDA or designated as not covered in the Coverage Issues Appendix. All drug determinations in the Coverage Issues Appendix and elsewhere in the manual are deleted, except those which must continue to be treated as exceptions.
>
> Under the new policy contractors will be responsible for determining whether or not FDA has approved drugs or biologicals, and we no longer will routinely issue national instructions concerning FDA approval of new drugs or biologicals, or new uses of drugs and biologicals. . . .
>
> Drug treatment must still be determined reasonable and necessary in individual cases . . . etc.

The package drugs used in Dr. (your doctor) chelation therapy are all or some of the following: disodium edetate, vitamin C, magnesium, heparin, vitamin B_6, and procaine. These may be followed, at the end of the infusion, by an intravenous injection of adrenal cortical extract, hydrochloric acid, pyridoxine, vitamin B_{12},and multivitamins.

According to Henry Herir, Director of Coverage Policy and Medical Services (room 463, East Bldg., Hi-Rise, 6401 Security Blvd., Baltimore, MD 21235), the previous exclusion against payment by Medicare for the use of disodium edetate (EDTA) in the treatment of atherosclerosis/arteriosclerosis was rescinded effective May 15, 1980. This change came about as part of a complete evaluation of Medicare drug policy.

In my own case, chelation therapy with disodium edetate or ethylene diamine tetraacetic acid (EDTA) is medically necessary. At the age of ________, I am facing (describe your health problem). The diagnosis of this health problem is (name the diagnosis).

Dr. (you doctor) has recommended EDTA chelation therapy as the means for me to be relieved of my health problem. My doctor or I will be happy to supply data including clinical history I've experienced over a length of time to justify the use of chelation therapy as essential to my well-being. He proposes a series of (number) treatments with EDTA.

In view of the change in policy taken by the Health Care Financing Administration, U.S. Dept. of HEW, SECTION 2050.5—*Drugs and Biologicals,* and the *Coverage Issues Appendix,* regarding drugs and biologicals and whereas Section 1801 of the Medicare Act provides that no Federal officer or employee shall exercise any supervision or control over the practice of medicine or the manner in which medical services are provided, and whereas my physician, Dr. (your doctor), after appropriate, accepted and approved testing methods, has determined my need for EDTA chelation therapy for my cardiovascular problem, I respectfully ask your permission and assurance of reimbursement for appropriate payment of fees for the services I propose to receive from my doctor.

Thank you for your quick response to my appeal. I am in desperate need of this treatment, a chemical endarterectomy, and solicit your consideration.

Yours sincerely,

Your name

Notes

Chapter 1

1. Elizabeth Rasche Gonzalez, "Constricting arteries expand views of ischemic heart disease." *Medical News, JAMA*, 243:4 (January 25, 1980), 309–16.

2. *Ischemic Heart Disease: The Role of Coronary Artery Spasm.* New York: Science and Medicine Publishing Co., Inc., 1979.

3. Johan Bjorksten, "Possibilities and limitations of chelation as a means for life extension." *Rejuvenation*, 8:3 (September 1980), 67–72.

Chapter 2

1. B. Greg Brown, "Coronary Vasospasm." *Arch. Inters. Med.*, 141 (May 1981), 716–22.

2. C. P. Lamar, "Chelation endarterectomy for occlusive atherosclerosis." *J. Amer. Geriatrics Soc.*, 14:3 (1966), 272–94.

3. Howard Wolinsky, "Area man: treatment restored vision." *Today*, January 9, 1977.

4. H. Rudolph Alsleben and Wilfrid E. Shute, *How to Survive the New Health Catastrophes.* Anaheim, California: Survival Publications, Inc., 1973.

Chapter 3

1. Robert W. Wissler and Draga Vesselinovitch, *Modern Concepts of Cardiovascular Disease.* American Heart Association, 1977.

2. Frank H. Netter, *The Ciba Collection of Medical Illustrations.* Vol. 5: *Heart.* Edited by Frederick I. Yonkman. Summit, N.J.: Ciba Pharmaceutical Company, 1969.

3. Henry J. Speedby, *The 20th Century and Your Heart.* Westport, Connecticut: Associated Booksellers, 1961.

4. "Constructing arteries." *Medical News, JAMA*, January 20, 1980.

5. Hans Selye, *Calciphylaxis.* Chicago: University of Chicago Press, 1956.

6. Harvey Wolinsky, "A new look at atherosclerosis." *Cardiovascular Medicine*, September 1976, pp. 41–54.

7. Johan Bjorksten, "Recent developments in protein chemistry." *Chemical Industries*, 48 (June 1941), 746–51.

8. Johan Bjorksten, "Chemistry of Duplication." *Chemical Industries*, 50 (January 1942), 68–72.

9. Garry F. Gordon and Robert B. Vance, "EDTA chelation therapy for arteriosclerosis: history and mechanisms of action." *Osteopathic Annals*, 4 (February 1976), 38–62.

10. R. G. Petersdorf, "Internal medicine in family practice." *New England J. Med.*, 295 (1975), 326.

11. D. H. Blankenhorn and L. E. Bolwock, "A quantitive study of coronary artery calcification." *Am. J. Pathology*, 39 (1961), 511.

12. A. Soffer, *Chelation Therapy*. Springfield, Illinois: Charles G. Thomas, 1964.

13. A. Soffer, T. Toribara, and A. Sayman, "Myocardial responses to chelation." *Br. Heart J.*, 23 (1961), 690–94.

14. A. Soffer, et al., "Clinical applications and untoward reactions of chelation in cardiac arrhythmia." *Arch. Intern. Med.*, 106 (1960), 824–34.

15. A. Popovivi, et al., "Experimental control of serum calcium levels *in vivo*." Georgetown University Medical Center, *Proc. Soc. Exp. Biol. Med.*, 74 (1950), 415–17.

16. M. Rubin, "Fifth conference on metabolic interrelations." *Biologic Action of Chelating Agents*, Macy Foundation, 1954, pp. 344–58.

17. A. B. Hastings, "Studies on the effect of alteration of calcium in circulating fluids on the mobility of calcium." *Trans. Macy Conf. Metabolic Interrelations* 3 (1951), 38–50.

18. C. P. Lamar, "Calcium chelation of atherosclerosis—nine years' clinical experience." Fourteenth Annual Meeting, American College of Angiology, 1968.

19. C. P. Lamar, "Chelation endarterectomy for occlusive atherosclerosis." *J. Am. Geriatr. Soc.*, 14 (1966), 272.

20. C. P. Lamar, "Chelation therapy of occlusive arteriosclerosis in diabetic patients." *Angiology*, 15 (1965), 379.

21. Robert F. Zelis, "Calcium entry blockers in cardiologic therapy." *Hospital Practice* 16 (August 1981), 49–56.

22. Robert F. Zelis and S. F. Flaim, "Calcium influx blockers and vascular smooth muscle: Do we really understand the mechanisms?" *Ann. Intern. Med.*, 94 (1981), 124.

23. E. M. Antman, P. H. Stone, J. E. Muller, and E. Braunwald, "Calcium channel blocking agents in the treatment of cardiovascular disorders. Part I: Basic and clinical electrophysiologic effects." *Ann. Intern. Med.*, 93 (1980), 875; P. H. Stone, E. M. Antman, J. E. Muller, and E. Braunwald, "Part II: Hemodynamic effects and clinical applications." Ibid. 886.

24. R. F. Zelis, A. J. Liedtke, and S. F. Flaim, "Antianginal drugs." In *Clinical Applications of Cardiovascular Drugs*, L. S. Dreifus and A. N. Brest, eds., Chapter 4. The Hague: Martinus Nijhoff, 1980.

25. R. F. Zelis and J. S. Schroeder, eds., "Calcium, calcium antagonists, and cardiovascular disease." *Chest*, 78 (Suppl.1, 1980), 121.

26. E. M. Antman, et al., "Nifedipine therapy for coronary-artery spasm." *N. Engl. J. Med.*, 302 (1980), 1269.

27. S. J. Rosenthal, et al., "The efficacy of diltiazem on coronary artery spasm." *Am. J. Cardiol*, 46 (1980), 1027.

28. Andrew M. Sincock, "Life extension in the rotifer by chelating agents." *J. Gerontology*, 50 (1975), 289-93.

Chapter 4

1. "Coronary arteriography and coronary artery surgery." *The Medical Letter*, 18, 14 (issue 456), July 2, 1976.

2. D. F. Adams, et al. *Circulation*, 48 (1973), 609.

3. Arthur Selzek, "Conventional measures of blood flows are inaccurate." *Medical World News*, December 22, 1980, p. 10.

4. L. M. Zir, et al. *Circulation* 53 (April 1976), 627.

5. *Medical World News*, December 24, 1979, p. 6.

6. E. D. Mundth and W. G. Austen, *New England Journal of Medicine*, 293 (1975), 13, 75, 124.

7. W. S. Fields, R. R. North, W. K. Hass, et al., "Joint study of extracranial arterial occlusion as a cause of stroke. I. Organization of study and survey of patient population." *Journal of the American Medical Association*, 203 (1968), 955.

8. C. M. Fisher, I. Gore, N. Okabe, et al., "Atherosclerosis of the carotid and vertebral arteries—extracranial and intracranial." *Journal of Neuropathology and Experimental Neurology*, 24 (1965), 455.

9. H. Hager, "Differential diagnosis of apoplexy by ophthalmodynamograph." *Triangle*, 6 (1964), 259.

10. M. H. Thomas and M. A. Petrohelas, "Diagnostic significance of retinal artery pressure in internal carotid involvement." *American Journal of Ophthalmology*, 36 (1953), 335.

11. M. M. Kartchner and L. P. McRae, "Auscultation for carotid bruits in cerebrovascular insufficiency." *JAMA*, 210 (1969), 494.

12. B. L. Segal, W. Likoff, and J. H. Moyer, *The Theory and Practice of Auscultation*. Philadelphia: F. A. Davis Co., 1964, pp. 164–269.

13. I. N. Kalb, "Retinal photography in an automated multiphasic screening program." *Proceedings of the Twentieth International Congress on Ophthalmology*, 2 (1966), 1028.

14. Lawrence K. Altman, "New method helps doctors expand the use of ultrasound in diagnosis." *New York Times*, October 20, 1977.

15. T. J. Ryan, et al., "Value of exercise ECG for diagnosing coronary disease." *J. Cardiovascular Medicine*, January 1980, pp. 61–77.

16. L. F. Vitale, et al., "Blood lead: an inadequate measure of occupational exposure." *J. Occupational Medicine*, 17:3 (1975), 155–56.

17. R. Jacob, et al., "Hair as a biopsy material. V. Hair metal as an index of hepatic metal in rats: copper and zinc." *American J. Clinical Nutrition*, 31:3 (March 1978), 477–80.

18. A. Prasad, *Trace Elements in Human Health and Disease*, Vols. I & II. New York: Academic Press, 1976.

19. D. R. Williams, ed., *An Introduction to Bio-Inorganic Chemistry*. Springfield, Ill.: Charles C. Thomas, 1976, p. 315.

20. A. Chattopadyay, et al., "Scalp hair as a monitor of community exposure to lead." *Archives of Environmental Health*, 32:5 (September-October 1977), 226–36.

Chapter 5

1. A. Graham and F. Graham, "Lead poisoning and the suburban child." *Today's Health*, March 1974.

2. R. J. Caprio, H. I. Margulis, and M. Joselow, "Lead absorption in children

and its relationship to urban traffic densities." *Arch. Environ. Health*, 28 (1975), 195–97.

3. R. Freeman, "Reversible myocarditis due to chronic lead poisoning in childhood." *Arch. Dis. Child.*, 40 (1965), 389–93.

4. A. M. Seppalainen et al., "Subclinical neuropathy at 'safe' levels of exposure." *Arch. Environ. Health*, 1975.

5. W. H. Strain et al., eds., *Clinical Application of Zinc Metabolism*. Springfield, Ill.: Charles C. Thomas, 1974.

6. J. P. Isaacs and J. C. Lamb, "Trace metals, vitamins, and hormones in ten-year treatment of coronary atherosclerotic heart disease." Library of Congress Catalog Card Number 74-82882, as delivered at the Texas Heart Institute Symposium on Coronary Artery Medicine and Surgery, Houston, February 21, 1974.

7. L. Marsh and F. C. Fraser, "Chelating agents and teratogenesis." *Lancet* (1973), 846.

8. H. M. Perry, Jr., and H. A. Schroeder, "Lesions resembling vitamin B-complex deficiency and urinary loss of zinc produced by EDTA." *Am. J. Med.*, 22 (1957), 168–72.

9. H. Foreman, "Toxic side effects of EDTA." *J. Chron. Dis.*, 16 (1963), 319–23.

10. H. M. Perry, Jr., and E. F. Perry, "Normal concentrations of some trace metals in human urine, changes produced by EDTA." *J. Clin. Invest.*, 38 (1959), 1452.

11. C. S. Tidball, "Nonspecificity of cation depletion when using chelators in biological systems." *Gastroenterology*, 60 (1971), 481.

12. J. T. Davies, *The Clinical Significance of the Essential Biological Metals*. Springfield, Ill.: Charles C. Thomas, 1972.

13. G. Schettler and A. Weizel, eds., *Atherosclerosis*. III: *Proceedings of the Third International Symposium*. New York, Heidelberg, Berlin: Springer Verlag, 1974.

14. N. Baumslag et al., "Hair-metal binding." *Environ. Health Perspec.* 8 (1974), 191–99.

15. T. A. Hinners et al., "Trace element nutriture and metabolism through head hair analysis." Department of Surgery, Case Western Reserve University School of Medicine, Cleveland, 1976.

16. D. I. Hammer et al., "Trace metals in human hair as a simple epidemiologic monitor of environmental exposure." National Environmental Research Center, Environmental Protection Agency, Research Triangle Park, N.C., 1977.

17. W. G. Hoekstra and J. W. Suttie, eds., *Trace Element Metabolism in Animals, II*. University Park Press, 1974.

18. V. N. Kurliandchikov, "Treatment of patients with coronary arteriosclerosis with Unithiol in combination with decamevit." *Vrach. Delo.*, 6 (1974), 8.

19. V. I. Zapadnick et al., "Pharmacological activity of Unithiol and its use in clinical practice." *Vrach. Delo.*, 8 (1973), 122.

20. O. Brucknerova and J. Tulacek, "Chelates in the treatment of occlusive atherosclerosis." *Unitr. Leg.*, 18 (1972), 729.

21. R. I. Levy et al., "Dietary and drug treatment of primary hyperlipoproteinemia." *Ann. Intern. Med*, 77 (1972), 267.

22. M. Walker, *Chelation Therapy: How to Prevent or Reverse Hardening of the Arteries*. Atlanta: '76 Press, 1980.

23. A. Popovici et al., "Experimental control of serum calcium levels *in vivo*."

Georgetown University Medical Center, *Proc. Soc. Ex. Biol. & Med.*, 74 (1950), 415–17.

24. A. Rostenberg, Jr., and A. J. Perkins, "Nickel and cobalt dermatitis." *J. Allergy*, 22 (1951), 466.

25. F. Proescher, "Anticoagulant Properties of Ethylene-diamine tetra-acetic acid." *Proc. Soc. Exper. Biol. & Med.*, 76 (1951), 619.

26. N. E. Clarke, C. N. Clarke, and R. D. Mosher, "The *in vivo* dissolution of metastatic calcium: an approach to atherosclerosis." *Am. J. Med. Sci.*, 229 (1955) 142.

27. N. E. Clarke, C. N. Clarke, and R. E. Mosher, "Treatment of angina pectoris with disodium-ethylenediaminetetraacetic acid." *Am. J. Med. Sci.*, 232 (1956), 654–66.

28. J. Cornfield and S. Mitchell, "Selected risk factors in coronary disease." *Arch. Environ. Health*, 19 (1969), 382–394.

29. M. Vavrik, "High risk factors and atherosclerotic cardiovascular diseases in the aged." *J. Am. Geriatr. Soc.*, 22 (1974), 203.

30. H. Malmros, "Primary dietary prevention of atherosclerosis." *Bibl. Nutr. Dieta*, 19 (1973), 108.

31. R. McGandy et al., "Dietary fats, carbohydrates and atherosclerotic vascular disease." *N. Engl. J. Med.*, 277 (1967), 186.

32. C. Moses, *Atherosclerosis: Mechanisms as a Guide to Prevention.* Philadelphia: Lea & Febiger, 1963.

33. National Academy of Sciences, National Research Council, *Dietary Fat and Human Health.* Publ. No. 1147, 1966.

34. O. A. Larsen and N. A. Lassen, "Effects of daily muscular exercise in patients with intermittent claudication." *Lancet*, 2 (1966), 1093.

35. H. Richard Casdorph, "EDTA chelation therapy, efficacy in arteriosclerotic heart disease." *Journal of Holistic Medicine*, 3:1 (Spring/Summer 1981), 53–59.

36. H. Richard Casdorph, "EDTA chelation therapy II, efficacy in brain disorders." *Journal of Holistic Medicine*, 3:2 (Fall/Winter 1981).

37. J. R. Kitchell et al., "The treatment of coronary artery disease with disodium EDTA—a reappraisal." *Am. J. Cardiol.* 11:501-506, 1963.

38. P. D. Doolan, et al., "An evaluation of the nephrotoxicity of ethylenediaminetetraacetate and diethylenetriaminepentaacetate in the rat." *Toxicology and Applied Pharmacology* 10:481–500, 1967.

Chapter 6

1. Bruce W. Halstead, *The Scientific Basis of EDTA Chelation Therapy.* Colton, California: Golden Quill Publishers, Inc., 1979, p. 85.

2. A. Catsche and A. E. Harmuth-Hoene, "Pharmacology and therapeutic applications of agents used in heavy metal poisoning." *Pharmac. Ther. Annals*, 1 (1978), 1–118.

3. B. Oser et al., "Safety evaluation studies of calcium EDTA." *Toxicol. Appl. Pharmacol.*, 5 (1963), 142–62.

4. L. S. Goodman and A. Gilman, *The Pharmacological Basis of Therapeutics*, 3rd ed. New York: Macmillan Co., 1966, p. 1785.

5. H. Foreman, C. Finnegan, and C. C. Lushbaugh, "Nephrotoxic hazard from uncontrolled EDTA calcium-disodium therapy." *JAMA*, 160 (1956) 1042–46.

6. H. R. Dudley and A. C. Ritchie, "Pathologic changes associated with the use of sodium EDTA in treatment of hypocalcemia." *N. Engl. J. Med.*, 252 (1955), 331–37.

7. M. J. Seven, "Observations on the dosage of I.V. chelating agents." *Antibiot. Med.*, 5 (1958), 251.

8. P. D. Doolan et al., "An evaluation of the nephrotoxicity of ethylenediaminetetraacetate and diethylenetriaminepentaacetate in the rat." *Toxicol. Appl. Pharmacol.*, 10 (1967), 481–500.

9. S. L. Schwartz, C. B. Johnson, and P. D. Doolan, "Study of the mechanism of renal vacuologenesis induced in the rat by EDTA." *Mol. Pharacol.*, 6 (1970), 54–60.

10. S. L. Schwartz et al., "Subcellular localization of EDTA in the proximal tubular cell of the rat kidney." *Biochem. Pharmacol.*, 16 (1967), 2413–19.

Chapter 7

1. Rollin H. Kimball, "The diet that can reverse heart disease." *New York*, December 11, 1978, pp. 77–86.

2. H. A. Guthrie, *Introductory Nutrition*. St. Louis: C. V. Mosby Co., 1975, pp. 11–12.

3. H. Ashmead et al., *J. Applied Nutrition*, V:26 (Summer 1974), 5.

4. H. Ashmead, "Mineral chelation means better health." *Let's Live*, 1975.

5. Johan Bjorksten, *Longevity, A Quest*. Madison, Wis.: Bjorksten Research Foundation, 1981, p. 164.

6. B. Saltin and J. Karlsson, *Muscle Metabolism During Exercise*. New York: Plenum Publishing Co., 1971, p. 395.

7. Johan Bjorksten, "The crosslinkage theory of aging as a predictive indicator." *Rejuvenation*, VIII:62 (September 1980).

8. D. B. Zilversmit and C. W. Adams, *Atherosclerosis*, edited by R. J. Jones, New York: Springer Verlag, 1970, p. 28–35.

9. C. E. Butterworth and C. Krumdieck, *American Journal of Clinical Nutrition*, August 1974.

10. Earl P. Benditt, "The origin of atherosclerosis." *Scientific American* (February 1977), pp. 74–82.

11. P. Handler, M.L.C. Bernheim, and J. R. Klein, *Journal of Biological Chemistry* 138 (1941), 211.

12. Hans W. Nieper, International Society for Infarct Control, *Circular Letter*, 28, October 10, 1977.

13. V. N. Kurliandihikov, "Treatment of patients with coronary arteriosclerosis with unithol in combination with Decamevit." *Vrach. Delo.* 6:8, 1973.

Chapter 8

1. W. Blumen, "Leaded gasoline and cancer mortality." *Scheiz Med. Wech.*, 106 (1976), 503–506.

2. C. F. Peng, J. J. Kane, M. L. Murphy, and K. D. Straub, "Abnormal mitochondrial oxydative phosphorylative of ischemic myocardium reversed by Ca^{++} -chelating agents." *J. Mol. Chem. Cardiol.*, 9 (1977), 897–908.

3. L. Triner et al., "Cyclic phosphodiestasterase activity and the action of papaverine." *Biochem. Biophys. Res. Commun.*, 40 (1970), 64–69.

4. P. Greengard et al., *Advances in Cyclic Nucleotide Research,* Volume 1. New York: Raven Press, 1972, pp. 195–211.

5. K. I. Sivjakov, "The treatment of acute selenium, cadmium and tungsten intoxication in rats with calcium disodium ethylenediaminetetraacetate." *Toxicol. Appl. Pharmacol.*, 1 (1959), 602–608.

6. J. Tessinger, "Biochemical responses to provocative chelation by edatate disodium calcium." *Arch. Environ. Health*, 23 (1971), 280.

7. L. M. Klevay, "Coronary heart disease: zinc/copper hypotheses." *Am. J. Clin. Nutr.*, 28 (1975), 764–74.

8. H. A. Schroeder, *The Poisons Around Us.* Bloomington, Ind.: Indiana University Press, 1974.

9. R. Freemen, "Reversible myocarditis due to chronic lead poisoning in childhood." *Arch. Dis. Child*, 40 (1965), 389–93.

10. J. M. Price, "Some effects of chelating agents on trypotophan metabolism in man." *Fed. Proc.* (Suppl. 10, 1961), 223–26.

11. R. Zelis et al., "Effects of hyperlipoproteinemias and their treatment on the peripheral circulation." *J. Clin. Invest.*, 49 (1970), 1007.

12. H. A. Schroeder and H. M. Perry, Jr., "Antihypertensive effects of binding agents." *J. Lab. Clin. Med.*, 46 (1955), 416.

13. Yeh Yu Shin, "Cross-linking of elastin in human atherosclerotic aortas." *Lab. Invest.*, 25 (1971), 121.

14. D. A. Hall, "Coordinately bound calcium as a cross-linking agent in elastin and as an activator of elastolysis." *Gerontologia*, 16 (1970), 325–39.

15. D. A. Wilder, "Mobilization of atherosclerotic plaque calcium with EDTA utilizing the isolation-persuasion principle." *Surgery*, 52 (1962), 5.

16. G. Schreiber, "*In vivo* thinning of thickened capillary basement membranes by rapid chelation." (Unpublished manuscript.)

17. A. J. Boyle, R. E. Mosher, and D. S. McCann, "Some vivo effects of chelation. 1: Rheumatoid arthritis." *J. Chronic Dis.*, 16 (1963), 325–28.

18. L. H. Leipzig, A. J. Boyle, and D. S. McCann, "Case histories of rheumatoid arthritis treated with sodium or magnesium EDTA." *J. Chronic Dis.*, 22 (1970), 553–63.

19. H.S.M. Uhl et al., "Effect of EDTA on cholesterol metabolism in rabbits." *Am. J. Clin. Pathol.*, 23 (1953), 1226–33.

20. H. S. Jacob, "Pathologic states of erythrocyte membrane." University of Minnesota, *Hospital Practice*, December 1974, pp. 47–49.

21. A. Soffer, T. Toribara, and A. Sayman, "Myocardial response to chelation." *Br. Heart J.*, 23 (1961), 690–94.

22. A. Soffer et al., "Clinical applications and untoward reactions of chelation in cardiac arrhythmias." *Arch. Intern. Med.*, 166 (1960), 824–34.

23. H. M. Perry, Jr., "Hypertension and the geochemical environment." *Ann. N.Y. Acad. Sci.*, 199 (1972), 202–228.

24. A. Popovici et al., "Experimental control of serum calcium levels *in vivo*." Georgetown University Medical Center, *Proc. Soc. Exp. Biol. Med.*, 74 (1950), 415–17.

25. C. P. Lamar, "Calcium chelation of atherosclerosis—nine years; clinical experience." Fourteenth Annual Meeting, American College of Angiology, 1968.

26. L. E. Meltzer, M. E. Urol, and J. R. Kitchell, "The treatment of coronary artery disease with disodium EDTA." In M. J. Seven, ed., *Metal Binding in Medicine*. Philadelphia: J. B. Lippincott Co., 1960, pp. 132–36.

27. L. E. Meltzer et al., "The urinary excretion pattern of trace metals in diabetes mellitus." *Am. J. Med. Sci.*, 244 (1962), 282–89.

28. J. R. Kitchell, L. E. Meltzer, and M. J. Seven, "Potential uses of chelation methods in the treatment of cardiovascular diseases." *Prog. Cardiovasc. Dis.*, 19 (1961), 798.

29. J. Bjorksten, "Crosslinking and the aging process." In M. Rockstein, ed., *Theoretical Aspects of Aging*. New York: Academic Press, 1974.

30. A. Koen, D. S. McCann, and A. J. Boyle, "Some *in vivo* effects of chelation. II: Animal experimentation." *J. Chronic Dis.*, 16 (1963), 329–33.

31. J. R. Kitchell, L. E. Meltzer, and E. Rutman, "Effects of ions on *in vitro* gluconeogenesis in rat kidney cortex slices." *Am. J. Physiol.*, 208 (1965), 841–46.

32. W. H. Strain et al., eds., *Clinical Application of Zinc Metabolism*. Springfield, Ill.: Charles C. Thomas, 1974.

33. M. Friedman, *Pathogenesis of Coronary Artery Disease*. San Francisco: McGraw-Hill, 1969.

34. B. Zohman, "Emotional factors in coronary disease." *Geriatrics*, 28 (1973), 110.

35. C. Miller, in W. H. Strain et al., eds., *Clinical Application of Zinc Metabolism*. Springfield, Ill.: Charles C. Thomas, 1974.

36. H. A. Peters, "Trace minerals, chelating agents and the porphyrias." *Fed. Proc.* (Suppl. 10, 1961), 227–34.

37. A. Timmerman and G. Kallistatos, "Modern aspects of chemical dissolution of human renal calculi by irrigation." *J. Urol.*, 95 (1966), 469–75.

38. R. E. Birk and C. E. Rupe, "The teatment of systemic sclerosis with EDTA, pyridoxine and resperpine." *Henry Ford Med. Bull*, 14 (June 1966), 109–18.

39. A. Sincock, "Life extension in the rotifer by application of chelating agents." *J. Gerontol.*, 30 (1975), 289–93.

40. C. P. Lamar, "Chelation therapy of occlusive arteriosclerosis in diabetic patients." *Angiology*, 15 (1964), 379.

Chapter 9

1. Justus J. Schifferes and Louis J. Peterson, *Essentials of Healthier Living*. New York: John Wiley & Sons, Inc., 1972.

2. Arthur Blumenfeld. *Heart Attack: Are You a Candidate?* New York: Pyramid Books, 1971.

3. "Coronary surgery assailed." *Associated Press* dispatch, February 5, 1981.

4. Marcia Millman, *The Unkindest Cut: Life in the Backrooms of Medicine*. New York: William Morrow and Co., Inc., 1977.

5. "Conference on bypass finds fiscal strain bearable." *Medical World News*, May 25, 1981, p. 26.

6. Jane E. Brody, "Doctors query bypass surgery as aid to heart." *New York Times*, November 22, 1976.

7. "Coronary bypass surgery is now safer." *American Family Practice*, 23 (May 1981), 226.

8. Jeanie Esajian, "UMC sued over heart surgery." *Sacramento Union*, June 18, 1981, p. A1.

9. A. S. Geha et al., *J. Thorac. Cardiovasc. Surg.*, 70 (1975), 414.

10. F. D. Loop et al., *Am. J. Cardiol.*, 37 (May 1976), 870.

11. P. L. Tecklenberg et al., *Circulation*, 52, Suppl. 1 (1975), 61.

12. R. A. Winkle et al., *Circulation*, 52, Suppl. 1 (1975), 61.

13. R. Berg, Jr., et al., *J. Thorac. Cardiovasc. Surg.*, 70 (1975), 432.

14. L. C. Winterscherd et al., *Amer. Surg.*, 41 (September 1975), 41.

15. W. J. Keon et al., *Can. Med. Assoc. J.*, 114 (February 21, 1976), 312.

16. E. L. Alderman et al., *N. Engl. J. Med.*, 288 (1973), 535.

17. M. M. Bassan et al., *N. Engl. J. Med.*, 290 (1974), 349.

18. M. V. Cohen and R. Gorlin, *Circulation*, 52 (1975), 275.

19. T. Takan et al., *Circulation*, 52, Suppl. II (1975), 143.

20. R. Selden et al., *N. Engl. J. Med.*, 293 (1975), 1329.

21. "Heart bypasses are often unnecessary, study says." *New York Times*, March 10, 1977.

22. "Coronary arteriography and coronary artery surgery." *The Medical Letter*, 18 (issue 456, July 2, 1976), 14.

23. Daniel Weisz, Robert I. Hamby, Julius W. Garvey, Allan Wolpowitz, and B. George Wisoff, "What causes recurrent angina after bypass surgery?" *New York State Journal of Medicine*, May 1981, p. 924.

24. "Bypass blocked—but still no angina." *Medical World News*, August 9, 1976, p. 68.

25. "Comparing treatments for angina." *Medical World News*, April 7, 1975.

26. R. L. Fulton and W. R. Blakely, "Lumbar sympathectomy: A procedure of questionable value in the treatment of arteriosclerosis obliteraus of the legs." *Am. J. Surg.*, 116 (1968), 735.

27. C. F. Peng, J. J. Kane, M. L. Murphy, and K. D. Straub, "Abnormal mitochondrial oxidative phosphorylation of ischemic myocardium reversed by Ca^{++}-chelating agents." *Journal of Molecular and Cellular Cardiology*, 9 (1977), 897–908.

28. Sidney Alexander, M.D. (head of the Cardiovascular Section, Lahey Clinic Medical Center, Burlington, Massachusetts), "The slow channel: Role of the calcium ion in cardiac conduction and muscle function." Presented to the American Academy of Medical Preventics, Atlanta, Georgia, May 9, 1981.

Chapter 10

1. "Michigan MDs lose federal antitrust ruling." *AMA News*, July 10, 1981, p. 1.

Chapter 11

1. "Report cites a surgeon in high mortality rate." *The New York Times*, October 4, 1981.

2. "U.S. faces malpractice payment for surgeon." *The New York Times*, February 7, 1982, p. 41.

3. William D. Carey, "Science policy and Congress." *Science*, 212 (May 15, 1981), 2.

4. Harry Schwartz, "The AMA isn't what it used to be." *Wall Street Journal*, March 12, 1981, p. 60.

5. "Court upholds FTC's 'conspiracy' slap of AMA." *Metabolic Reporter*, 1 (January 1981), 1.

6. "FTC warns medicine: don't play monopoly." *Medical World News*, August 3, 1981, pp. 6 and 7.

7. James H. Scheuer, "The FDA too slow." *New York Times*, May 22, 1980.

8. "Probers call for FDA to end secrecy." *The Advocate*, June 1, 1977, p. 21.

Bibliographic Key

For reasons of space, the Bibliography uses standard abbreviations for the journals cited. This key lists these abbreviations and the full names of the journals. Readers who wish to investigate the Notes and Bibliography further can use a number of references available at their library, including the National Library of Medicine's *List of Journals Indexed in Index Medicus* or the *Union List of Serials.*

Acta. Endocr.
Journal of Endocrinology

Acta. Haemat.
Journal of Haematology

Acta. Med. Scand.
Medical Journal of Scandinavia

Acta. Otolaryng.
Journal of Otolaryngology

Acta. Physiol. Scand.
Physiology Journal of Scandinavia

Am. Heart J.
American Heart Journal

Am. J. Anat.
American Journal of Anatomy

Am. J. Cardiol.
American Journal of Cardiology

Am. J. Clin. Path.
American Journal of Clinical Pathology

Am. J. Dis. Child
American Journal of Diseases of Children

Am. J. Gastroent.
American Journal of Gastroenterology

Am. J. Med. Sci.
American Journal of Medical Science

Am. J. Ophthal.
American Journal of Ophthalmology

Am. J. Physiol.
American Journal of Physiology

Am. J. Vet. Res.
American Journal of Veterinarian Research

AMA Arch. Derm.
American Medical Association Archives of Dermatology

AMA Arch. Intern. Med.
American Medical Association Archives of Internal Medicine

AMA Arch. Indust. Health
American Medical Association Archives of Industrial Health

Anal. Biochem.
Annals of Biochemistry

Anat. Rec.
Anatomy Record

Ann. Intern. Med.
Annals of Internal Medicine

Ann. NY Acad. Sci.
Annals of the New York Academy of Science

Ann. Otol.
Annals of Otolaryngology

Arch. Biochem. Biophys.
Archives of Biochemistry and Biophysics

Arch. Derm.
Archives of Dermatology

Arch. Dis. Child
Archives of Disease in Childhood

Arch. Environ. Health
Archives of Environmental Health

Arch. Inc. Pharmacodyn
Inclusive Archives of Pharmacodynamics

Arch. Int. Pharmacodyn Ther.
International Archives of Pharmacodynamics and Therapeutics

Arch. Oral. Biol.
Archives of Oral Biology

Arch. Path.
Archives of Pathology

Arthritis Rheum.
Arthritis and Rheumatism

Aust. Ann. Med.
Australian Annals of Medicine

Bibl. Haemat.
Bibliotheca Haematologica

Biochem. Biophys. Acta.
Biochemistry and Biophysics Journal

Biochem. J.
Biochemical Journal

Biochem. Pharmacol.
Biochemistry and Pharmacology

Biochem. Res. Commun.
Biochemistry and Biophysics Research Communication

Brit. Heart J.
British Heart Journal

Brit. J. Derm.
British Journal of Dermatology

Brit. J. Indust. Med.
British Journal of Industrial Medicine

Brit. J. Ophthal.
British Journal of Ophthalmology

C. R. Acad. Sci.
Canadian Royal Academy of Science

C. R. Soc. Biol.
Canadian Royal Society of Biology

Canad. J. Biochem.
Canadian Journal of Biochemistry

Canad. J. Microbiol.
Canadian Journal of Microbiology

Canad. J. Physiol. Pharmacol.
Canadian Journal of Physiology and Pharmacology

Canad. Med. Assoc. J.
Canadian Medical Association Journal

Chem-Biol. Interact
Chemico-Biological Interactions

Chem. Rev.
Chemical Review

Circ. Res.
Circulation Research

Clin. Chem.
Clinical Chemistry

Clin. Chem. Acta.
Clinical Chemistry Journal

Comp. Biochem. Physiol.
Comparative Biochemistry and Physiology

Deutsch Gesundh
German Health

Endocrinology

Environ. Intern.
Environment International

Europ. J. Biochem.
European Journal of Biochemistry

Exp. Cell. Res.
Experimental Cell Research

Fed. Proc.
Federation Proceedings

Gastroenterology

Guy Hosp. Rep.
Guy Hospital Report

Henry Ford Hosp. Med. Bull.
Henry Ford Hospital Medical Bulletin

Isr. J. Med. Sci.
Israeli Journal of Medical Science

JAMA
Journal of the Americal Medical Association

J. Am. Geriat. Soc.
Journal of the American Geriatric Society

J. Bact.
Journal of Bacteriology

J. Biochem.
Journal of Biochemistry

J. Biol. Chem.
Journal of Biology and Chemistry

J. Cell. Biol.
Journal of Cell Biology

J. Cell. Comp. Physiol.
Journal of Cellular and Comparative Physiology

J. Cell. Physiol.
Journal of Cellular Physiology

J. Chromatogr.
Journal of Chromatography

J. Chron. Dis.
Journal of Chronic Diseases

J. Clin. Endocr.
Journal of Clinical Endocrinology

J. Clin. Invest.
Journal of Clinical Investigation

J. Dent. Res.
Journal of Dental Restoration

J. Embryol. Exp. Morph.
Journal of Embryology and Experimental Morphology

J. Exp. Zool.
Journal of Experimental Zoology

J. Gen. Microbiol.
Journal of General Microbiology

J. Immun.
Journal of Immunology

J. Infect. Dis.
Journal of Infectious Diseases

J. Invest. Derm.
Journal of Investigative Dermatology

J. Lab. Clin. Med.
Journal of Laboratory and Clinical Medicine

J. Lipid Res.
Journal of Lipid Research

J. Med. Chem.
Journal of Medicinal Chemistry

J. Morph.
Journal of Morphology

J. Neurol. Neurosurg. Psychiat.
Journal of Neurology, Neurosurgery and Psychiatry

J. Nucl. Med.
Journal of Nuclear Medicine

J. Nutr.
Journal of Nutrition

J. Path.
Journal of Pathology

J. Pediat.
Journal of Pediatrics

J. Pharma. Sci.
Journal of Pharmaceutical Sciences

J. Pharmacol. Exp. Ther.
Journal of Pharmacology and Experimental Therapeutics

J. Physiol.
Journal of Physiology

J. Surg. Res.
Journal of Surgical Research

J. Ultrastruct. Res.
Journal of Ultrastructure Research

Jap. J. Exp. Med.
Japanese Journal of Experimental Medicine

Jap. J. Med. Sci. Biol.
Japanese Journal of Medical Science and Biology

Life Sci.
Life Sciences

Med. J. Aust.
Medical Journal of Australia

New Engl. J. Med.
New England Journal of Medicine

Oral Surg.
Oral Surgery

Presse Med
Medical Press

Proc. Natl. Acad. Sci., USA
Proceedings of the National Academy of Sciences, United States of America

Proc. Soc. Exp. Biol. Med.
Proceedings of the Society for Experimental Biology and Medicine

Proc. West. Pharmacol. Soc.
Proceedings of the Western Pharmacology Society

Progr. Cardiov. Dis.
Progress in Cardiovascular Diseases

Progr. Hematol.
Progress in Hematology

Recent Adv. Biol. Psychiat.
Recent Advances in Biological Psychiatry

Scand. J. Clin. Lab. Invest.
Scandinavian Journal of Clinical and Laboratory Investigation

Scand. J. Haemat
Scandinavian Journal of Haematology

Sci. Am.
Scientific American

South. Med. J.
Southern Medical Journal

Surg. Forum
Surgical Forum

Texas State J. Med.
Texas State Journal of Medicine

Throm. Diath. Haemorrh.
Thrombosis, Diathermy and Haemorrhage

Toxicol. Appl. Pharmacol.
Toxicology and Applied Pharmacology

Vnitr. Lek.
Vnitrni Lekarstvi

Vox Sang
Vox Sanguinis

West. J. Med.
Western Journal of Medicine

West. Med.
Western Medicine

Wisconsin Med. J.
Wisconsin Medical Journal

Yokohama Med. Bull.
Yokohama Medical Bulletin

Bibliography

The author index that follows represents approximately 20 percent of the existing bibliography up to June 1976 available through a scientific literature search for information about the therapeutic applications of chelation and/or ethylene diamine tetraacetic acid (EDTA). Because of the FDA and organized medicine's discouragement of chelation therapy usage for atherosclerosis, no research studies have been carried to completion since 1976 except for the five listed in the text (see Notes section). The number in parentheses following the citation indicates the number of scientific references cited in the paper.

Ahmed, S. S., Melgiri, S., and Abraham, G. J. S. "Potentiation of codeine analgesia by disodium edetate in albino rats." *Arch. Int. Pharmacodyn Ther.,* 195:357–60, 1972. (32)

Ahrens, F. A., and Aronson, A. L. "A comparative study of the toxic effects of calcium and chromium chelates of ethylenediaminetetraacetate in the dog." *Toxicol. Appl. Pharmacol.,* 18:10–25, 1971. (32)

Albahary, C., Renault, J., Labarbe, P., and Giraud, J. "The treatment of siderosis by desferrioxamine B. Comparison with the action of 2 other chelating agents." *Presse Med.,* 70:2823–26, 1962. (18)

Albert, A. "Design of chelating agents for selected biological activity." *Fed. Proc.,* 20 (Suppl. 10): 137–47, September 1961. (58)

Ali, R. and Evans, J. L. "Effect on dietary calcium, buffering capacity, lactose and EDTA on pH of a calcium absorption from gastrointestinal segments in the growing rat." *J. Nutr.*, 93:273-79, 1967. (30)

Altman, J., Wakim, K. G., and Winkelmann, R. K. "Effects of edathamil disodium on the kidney." *J. Invest. Derm.*, 38:215-18, 1962. (5)

Amar-Costesec, A. "Analytical study of rat liver microsomes treated by EDTA or pyrophosphate." *Arch. Int. Physiol. Biochim.*, 81:358-59, 1973. (4)

Amtorp, O. and Sorensen, S. C. "The permeability of the blood-brain barrier to ^{51} Cr-EDTA in rabbits with experimental allergic encephalomyelitis." *Acta. Neurol. Scand.*, 49:327-30, 1973. (9)

Andrews, B. F. "Calcium disodium edathamil therapy of lead intoxication. The significance of aminoaciduria." *Arch. Environ. Health,* 3:563-67, 1961. (27)

Andrews, B. F. "Hazards of edathamil (EDTA) therapy in lead intoxication." (Letters) *Pediatrics,* 28:161-62, 1961. (6)

Anghileri, L. J. "The binding of EDTA to human serum albumin." *Naturwisenschaften,* 55:182, 1968. (0)

Angle, C. R., and McIntire, M. S. "Lead poisoning during pregnancy. Fetal tolerance of calcium disodium edetate." *Am. J. Dis. Child,* 108:436-39, 1964. (24)

Anonymous. "The effects of chelating agents upon the atherosclerotic process." *Nutr. Rev.*, 21:352, 1963. (3)

Aposhian, H. V. "Biochemical and pharmacological properties of the metal-binding agent penicillamine." *Fed. Proc.,* 20 (Suppl. 10): 185-90, September 1961. (19)

Arcos, J. C., and Argus, M. F. "A reconsideration of the isolation in presence of EDTA and the swelling-contraction characteristics of rat heart sarcosomes. Metabolic topography of the sarcosomal contractile protein." *Biochemistry,* 3:2028-40, 1964. (60)

Aronov, D. M. "1st experience with the treatment of atherosclerosis patients with calcinosis of the arteries with trilon B (disodium salt of EDTA)." *Klin Med,* (Moskva) 41:19-23, May 1963. (15)

Aronson, A. L., Hammond, P. B., and Strafuss, A. C. "Studies with calcium ethylenediaminetetraacetate in calves; toxicity and use in bovine lead poisoning." *Toxicol. Appl. Pharmacol.,* 12:337-49, 1968. (26)

Aronson, A. L., and Rogerson, K. M. "Effect of calcium and chromium chelates of ethylenediaminetetraacetate on intestinal permeability and collagen metabolism in the rat." *Toxicol. Appl. Pharmacol.,* 21:440-53, 1972. (34)

Arvidson, S. "Studies on extracellular proteolytic enzymes from *Staphylococcus aureaus.* II. Isolation and characterization of an EDTA sensitive protease." *Biochim. Biophys. Acta.,* 302:149-57, 1973. (21)

Asano, M. "Some effects of disodium ethylenediamine tetraacetate on growth of transplantable mouse tumors." *Jap. J. Med. Sci. Biol.,* 12:365-74, 1959. (15)

Baetsle, L., and Bengsch, E. "Ion-exchange characteristics of the radium-ethylenediaminetetraacetate complex." *J. Chromatogr.,* 8:265-73, 1962. (10)

Belanger, L. F., Copp, D. H., and Morton, M. A. "Demineralization with EDTA by constant replacement." *Anat. Rec.,* 153:41-47, 1965. (20)

Bernheim, F. "The effect of ethylenediaminetetraacetate and its calcium chelate on the rate of swelling of a strain of Pseudomonas aeruginosa in salt solutions." *Canad. J. Microbiol.,* 18:1643-46, 1972. (9)

Bianchi, C. P. "The effect of EDTA and SCN on radiocalcium movement in frog rectus abdominis muscle during contractures induced by calcium removal." *J. Pharmacol. Exp. Ther.*, 147:360–70, 1965. (25)

Billups, C., Pape, L., and Saltman, P. "The kinetics and mechanism of Fe (III) exchange between chelates and transferrin. III. The amount of iron-ethylenediaminetetraacetate initially bound to protein." *J. Biol. Chem.*, 242:4284–86, 1967. (4)

Birk, R. E., and Rupe, C. E. "Systemic sclerosis. Fourteen cases treated with chelation (disodium EDTA) and/or pyridoxine, with comments on the possible role of altered tryptophan metabolism in pathogenesis." *Henry Ford Hosp. Med. Bull.*, 10:523–53, 1962. (54)

Blijenberg, B. G., and Leijnse, B. "A simple method for the determination of EDTA in serum and urine." *Clin. Chim. Acta.*, 26:577–79, 1969. (1)

Blumen, W., and Reich, Th. "Leaded gasoline—A cause of cancer." *Environment International*, 3:465–71, Pergamon Press, Ltd. (1980) Great Britain (62)

Botts, J., Chashin, A. and Schmidt, L. "Computation of metal binding in bi-metal—bi-chelate systems." *Biochemistry*, 5:1360–64, 1966. (5)

Botts, J., Chashin, A., and Young, H. L. "Alkali metal binding by ethylenediaminetetraacetate, adenosine 5'-triphosphate, and pyrophosphate." *Biochemistry*, 4:1788–96, 1965. (9)

Boyadjian, N. "Intensification of the effects of quinidine on experimental auricular fibrillation in the dog previously treated with the disodium salt of ethylenediaminetetraacetate." *C. R. Soc. Biol.*, 155:414–16, 1961. (6)

Boyle, A. J., Clarke, N. E., Mosher, R. E., and McCann, D. S. "Chelation therapy in circulatory and sclerosing diseases." *Fed. Proc.*, 20 (Suppl 10): 243–52, September 1961. (19)

Boyle, A. J., Mosher, R. E., and McCann, D. S. "Some in vivo effects of chelation. I. Rheumatoid arthritis." *J. Chron. Dis.*, 16:325–28, 1963. (5)

Brading, A. F., and Jones, A. W. "Distribution and kinetics of CoEDTA in smooth muscle, and its use as an extracellular marker." *J. Physiol.*, 200:387–401, 1969. (26)

Brien, T. G., and Fay, J. A. "^{51}Cr-EDTA biological half life as an index of renal function." (Letters) *J. Nucl. Med.*, 13:339–40, 1972. (4)

Brucknerova, O., and Tulacek, J. "Chelates in the treatment of occlusive arteriosclerosis." *Vnitr. Lek.*, 18:729–36, 1972. (25)

Brugsch, H. G. "Fatal nephropathy during edathamil therapy in lead poisoning." *AMA Arch. Indust. Health*, 20:285–92, 1959. (41)

Cassidy, M. M., and Tidball, C. S. "Cellular mechanism of intestinal permeability alterations produced by chelation depletion." *J. Cell. Biol.*, 32:685–98, 1967. (27)

Castellino, N., and Aloj, S. "Effects of calcium sodium ethylenediamine-tetra-acetate on the kinetics of distribution and excretion of lead in the rat." *Brit. J. Indust. Med.*, 22:172–80, 1965. (25)

Castronovo, Jr., F. P., Reba, R. C., and Wagner, Jr., H. N. "System for sustained intravenous infusion of a sterile solution of ^{137m}Ba-ethylenediaminetetraacetic acid (EDTA)." *J. Nucl. Med.*, 10:242–45, 1969. (10)

Chevance, L. G. "Biological study of a chelating agent with an affinity for calcium in the field of otology." *Acta Otolaryng.*, 51:46–54, 1960. (0)

Chisolm, Jr., J. J. "Chelation therapy in children with subclinical plumbism." *Pediatrics*, 53:441–43, 1974. (17)

Chisolm, Jr., J. J. "The use of chelating agents in the treatment of acute and chronic lead intoxication in childhood." *J. Pediat.,* 73:1–38, 1968. (92)

Chung, R. S. K., Sum, P. T., Goldman, H., Field, M., and Silen, W. "Effects of chelation of calcium on the gastric mucosa barrier." *Gastroenterology,* 59:200–7, 1970. (17)

Clark, W. G. "Hyperthermic effect of disodium edetate injected into the lateral cerebral ventricle of the unanesthetized cat." *C. R. Acad. Sci.,* 250:3536–38, 1960. (6)

Clarke, Sr., N. E. "Atherosclerosis, occlusive vascular disease and EDTA." (Ed) *Am. J. Cardiol.,* 6:233–36, 1960. (18)

Clarke, Sr., N. E., Clarke, Jr., N. E., and Mosher, R. E. "Treatment of occlusive vascular disease with disodium ethylenediamine tetraacetic acid (EDTA)." *Am. J. Med. Sci.,* 239:732–44, 1960. (50)

Cohen, P., Cooley, M. H., and Gardner, F. H. "Platelet preservation. III. Comparison of radioactivity yields of platelet concentrates derived from blood anticoagulated with EDTA and ACD." *New Eng. J. Med.,* 273:845–50, 1965. (16)

Cohen, S., Weissler, A. M., and Schoenfeld, C. D. "Antagonism of the contractile effect of digitalis by EDTA in the normal human ventricle." *Am. Heart J.,* 69:502–14, 1965. (20)

Conyers, R. A. J., Birkett, D. J., Neale, F. C., Posen, S., and Brudenell-Woods, J. "The action of EDTA on human alkaline phosphatase." *Biochim. Biophys. Acta.,* 139:363–71, 1967. (44)

Copp, D. H., Cheney, B. A. , and Stokoe, N. M. "Simple and precise micro-method for EDTA titration of calcium." *J. Lab. Clin. Med.,* 61:1029–37, 1963. (11)

Cottin, S., and Paupe, J. "Influence of a calcium chelating agent, ethylenediamine tetraacetic acid-disodium salt, on contractions of the isolated rabbit intestine." *C. R. Soc. Biol.,* 156:1587–90, 1962. (1)

Craven, P. C., and Morelli, H. F. "Chelation Therapy." *West. J. Med.,* 122:277–78, 1975. (14)

Crevier, M., and Dowd, G. F. "An electrocardiographic study of electrolyte-steroid-cardiopathy and of its prevention by disodium ethylenediamine tetraacetate." *Arch. Int. Pharmacodyn.,* 127:296–306, 1960. (21)

Curdel, A., Naslin, L., and Labeyrie, F. "Zinc reactivation of D-lactic dehydrogenase inhibited by ethylenediaminetetraacetic acid." *C. R. Acad. Sci.,* 249:1959–61, 1959. (7)

Curran, P. F., Zadunaisky, J., and Gill, Jr., J. R. "The effect of ethylenediaminetetraacetate on ion permeability of the isolated frog skin." *Biochim. Biophys. Acta.,* 52:392–95, 1961. (5)

Daniel, E. E., and Irwin, J. "On the mechanisms whereby EDTA, EGTA, DPTA, oxalate, desferrioxamine, and 1, 10-phenathroline affect contractility of rat uterus." *Canad. J. Physiol. Pharmacol.,* 43:111–36, 1965. (25)

Darwish, N. M., and Kratzer, F. H. "Metabolism of ethylenediaminetetraacetic acid (EDTA) by chickens." *J. Nutr.,* 86:187–92, 1965. (16)

Davis, F. A., Becker, F. O., Michael, J. A., and Sorensen, E. "Effect of intravenous sodium edetate (Na_2EDTA), and hyperventilation on visual and oculomotor signs in multiple sclerosis." *J. Neurol. Neurosurg. Psychiat.,* 33:723–32, 1970. (34)

Davis, H., and Moe, P.J. "Favorable response of calcinosis universalis to edathamil disodium." *Pediatrics,* 24:780-85, 1959. (18)

Davis, P. S., and Deller, D. J. "Effect of orally administered chelating agents EDTA, DTPA and fructose on radioiron absorption in man." *Aust. Ann. Med.*, 16:70–74, 1967. (15)

Dazord, A., Gallet, D., Bretrand, J., and Saez, J. M. "Action of calcium, EGTA and EDTA on adenyl cyclase activity in subcellular preparations of corticoadrenal glands." *C. R. Acad. Sci.*, 277D:1917–20, 1973. (9)

Dehaan, R. L. "The effects of the chelating agent ethylenediamine tetraacetic acid on cell adhesion in the slime mould *Dictyostelium discoideum*." *J. Embryol. Exp. Morph.*, 7:335–43, 1959. (32)

Dekel, N., Shahar, A., Soferman, N., and Kraicer, P. F. "Effect on EDTA on human cumulus granulosa cells." *Isr. J. Med. Sci.*, 8:2004–2007, 1972. (4)

DiPaolo, J. A. "SH-SS relationships in Ehrlich ascites tumor cells exposed to edathamil." *J. Cell. Comp. Physiol.*, 65:57–61, 1965. (14)

Dixit, P. K., and Lazarow, A. "EDTA (ethylene diamine tetraacetic acid) potentiation of insulin activity as assayed by the epididymal adipose tissue technique." *Proc. Soc. Exp. Biol. Med.*, 118:368–72, 1965. (13)

Dixon, J. S., Hunter, J. A. A., and Steven, F. S. "An electron microscopic study of the effect of crude bacterial -amylase and ethylenediaminetetraacetic acid on human tendon." *J. Ultrastruct. Res.*, 38:466–72, 1972. (12)

Donald, G. F., Hunter, G. A., Roman, W., and Taylor, A. E. J. "Current concepts of cutaneous porphyria and its treatment with particular reference to the use of sodium calcium-edetate." *Brit. J. Derm.*, 82:70–75, 1970. (24)

Doolan, P. D., Schwartz, S.L., Hayes, J. R., Mullen, J. C., and Cummings, N. B. "An evaluation of the nephrotoxicity of ethylenediaminetetraacetate and diethylenetriaminepentaacetate in the rat." *Toxicol. Appl. Pharmacol.*, 10:481–500, 1967. (36)

Drews, G. A., and Engel, W. K. "Reversal of the ATPase reaction in muscle fibres by EDTA. *Nature.*, 212:1551–53, 1966. (24)

Dudley, H. R., Ritchie, A. C., Schilling, A., and Baker, W. M. "Pathologic changes associated with the use of sodium ethylene diamine tetra-acetate in the treatment of hypercalcemia." *New Eng. J. Med.*, 252:331–37, 1955. (43)

Eliot, R. S., and Blount, Jr., S. G. "Calcium, chelates, and digitalis. A clinical study." *Am. Heart J.*, 62:7–21, 1961. (15)

Emmerson, B. T. "Chronic lead nephropathy. The diagnostic use of calcium EDTA and the association with gout." *Aust. Ann. Med.*, 12:310–24, 1963. (64)

Estep, H. L., Gardner, Jr., C. T., Taylor, J. P., Minott, A., and Tucker, Jr., H. St. G. "Phosphate excretion patterns following intravenous injection of ethylenediaminetetraacetate (EDTA)." *J. Clin. Endocr.*, 25:1385–92, 1965. (21)

Evans, P. J., and Eustace, P. "Scleromalacia perforans associated with Cron's disease treated with sodium versenate (EDTA)." *Brit. J. Ophthal.*, 57:330–35, 1973. (12)

Fabiny, R. J. "Effects of versene and B-2-thienylalanine on the developing down feather." *Am. J. Anat.*, 104:275–93, 1959. (13)

Feinstein, M. G. "Inhibition of muscle rigor by EDTA and EGTA." *Life Sci.*, 5:2177–86, 1966. (21)

Fink, C. W., and Baum, J. "Treatment of calcinosis universalis with chelating agents." *Am. J. Dis. Child.*, 105:390–92, 1963. (5)

Flowers, H. H., and Goucher, C. R. "The effect of EDTA on the extent of tissue damage caused by the venoms of *Bothrops atrox* and *Agkistrodon piscivorus.*" *Toxicon.*, 2:221–24, 1965. (7)

Flynn, D. M. "5-Year controlled trial of chelating agents in treatment of thalassaemia major." *Arch. Dis. Child.,* 48:829, 1973. (0)

Foreland, M., Pullman, T. N., Lavender, A. R., and Aho, I. "The renal excretion of ethylenediaminetetraacetate in the dog." *J. Pharmacol. Exp. Ther.,* 153:142–47, 1966. (20)

Foreman, H. "Summary remarks by the chairman. [Proceedings of a conference on biological aspects of metal-binding.]" *Fed. Proc.,* 20 (Suppl. 10): 257–58, September 1961. (0)

Foreman, H. "Toxic side effects of ethylenediaminetetraacetic acid." *J. Chron. Dis.,* 16:319–23, 1963. (20)

Foreman, H. "Use of chelating agents in treatment of metal poisoning (with special emphasis on lead)." *Fed. Proc.,* 20 (Suppl. 10): 191–96, September 1961. (80)

Foreman, H., Finnegan, C., and Lushbaugh, C. C. "Nephrotoxic hazard from uncontrolled edathamil calcium-disodium therapy." *JAMA,* 160:1042–46, 1956. (7)

Forssman, O., and Nordgvist, P. "The action in vitro and in vivo of sodium versenate on the phagocytic activity of neutrophile leukocytes." *Acta Haemat.,* 31:289–93, 1964. (9)

Fox, J. L., Higuchi, W. I., Fawzi, M., Hwu, R. C., and Hefferen, J. J. "Two-site model for human dental enamel dissolution in EDTA." *J. Dent. Res.,* 53:939, 1974. (4)

Foye, W. O. "Design of chelating agents for selected biological activity." *Fed. Proc.,* 20 (Suppl. 10): 147–49, September 1961. (18)

Foye, W. O., Solis, M. C. M., Schermerhorn, J. W., and Prien, E. L. "Effect of metal-complexing agents on mucopolysaccharide sulfate biosynthesis." *J. Pharm. Sci.,* 54:1365–67, 1965. (8)

Fried, J. F., Schubert, J., and Lindenbaum, A. "Action of edathamil (EDTA) analogs on experimental lead poisoning." *AMA Arch. Indust. Health,* 20:473–76, 1959. (22)

Friedel, W., Schulz, F. H., and Schroder, I. "Arteriosclerosis therapy with mucopolysaccharides and EDTA." *Deutsch Gesundh,* 20:1566–70, 1965. (37)

Fromke, V. L., Lee, M. Y., and Watson, C. J. "Porphyrin metabolism during Versente therapy in lead poisoning. Intoxication from an unusual source." *Ann. Intern. Med.,* 70:1007–12, 1969. (29)

Fuleihan, F. J. D., Kurban, A. K., Abboud, R. T., Beidas-Jubran, N., and Farah, F. S. "An objective evaluation of treatment of systemic scleroderma with disodium EDTA, pyridoxine and reserpine." *Brit. J. Derm.,* 80:184–89, 1968. (20)

Gesinski, R. M., and Morrison, J. H. "The effect of the anticoagulant EDTA on oxygen uptake by bone-marrow cells." *Experientia,* 24:296–97, 1968. (8)

Gillund, T. D., Howard, P. L., and Isham, B. "A serum agglutinating human red cells exposed to EDTA." *Vox Sang,* 23:369–70, 1972. (4)

Gingell, D., and Garrod, D. R. "Effect of EDTA on electrophoretic mobility of slime mould cells and its relationship to current theories of cell adhesion." *Nature,* 221:192–93, 1969. (12)

Godal, H. C. "The effect of EDTA on human fibrinogen and its significance for the coagulation of fibrinogen with thrombin." *Scand. J. Clin. Lab. Invest.,* 12 (Suppl. 53): 1–20, 1960. (32)

Goldner, A. M., Cassidy, M. M., and Tidball, C. S. "Nonspecificity of the divalent cation capable of restoring normal intestinal permeability after chelation depletion." *Proc. Soc. Exp. Biol. Med.,* 124:884–87, 1967. (16)

Goldschmidt, M. C., and Wyss, O. "Chelation effects on Azotobacter cells and cysts." *J. Bact.,* 91:120–24, 1966. (9)

Goldstein, M., Lauber, E., and McKereghan, M. R. "The inhibition of dopamine-B-hydroxylase by tropolone and other chelating agents." *Biochem. Pharmacol.,* 13:1103–06, 1964. (12)

Gorby, C. K., and Rieders, F. "The effect of simultaneous oral edathamil calcium disodium and lead acetate on lead accumulation in tissues of rats." *Arch. Inc. Pharmacodyn.,* 125:153–60, 1960. (12)

Gosselin, L. "The acylation of endogenous lysophospholipids in isolated rat liver microsomal fraction: influence of glutathione and ethylenediaminetetraacetate on this process." *Biochem. J.,* 130:43P, 1972. (2)

Goucher, C. R., and Flowers, H. H. "The chemical modification of necrogenic and proteolytic activities of venom and the use of EDTA to produce *Agkistrodon piscivorus,* a venom toxoid." *Toxicon.,* 2:139–47, 1964. (10)

Gould, R. G. "Metals and chelating agents in relation to atherosclerosis." *Fed. Proc.,* 20 (Suppl. 10): 252–53, September 1961. (4)

Graca, J. G., Davison, F. C., and Feavel, J. B. "Comparative toxicity of stable rare earth compounds. II. Effect of citrate and edetate complexing on acute toxicity in mice and guinea pigs." *Arch. Environ. Health,* 5:437–44, 1962. (5)

Graham, J. M., and Keatinge, W. R. "Sucrose-gap recording of prolonged electrical activity from arteries in Ca-free solution containing EDTA at low temperature." *J. Physiol.,* 208:2P–3P, 1970. (3)

Greenfield, S., and Klein, H. P. "Lipid synthesis in low protein homogenates of yeast. Effects of mitochondria and ethylenediaminetetraacetate." *J. Bact.,* 79:691–96, 1960. (11)

Grevisse, J. "Modification by ethylenediaminetetraacetic acid of the effects of lead ion on the isolated non-gravid uterus and jejunum of rabbits." *C. R. Soc. Biol.,* 153:702–704, 1959. (5)

Gunther, R. "Influence of chelating agents on the distribution and excretion of radioiron in rats." *Nauyn. Schmiedeberg. Arch. Pharm. Exp. Path.,* 262:405–418, 1969. (32)

Gustafson, R. L. "Formation of polynuclear chelates." *Fed Proc.,* 20 (Suppl. 10): 32–34, September 1961. (5)

Hammond, P. B. "The effects of chelating agents on the tissue distribution and excretion of lead." *Toxicol. Appl. Pharmacol.,* 18:296–310, 1971. (23)

Hammond, P. B., Aronson, A. L., and Olson, W. C. "The mechanism of mobilization of lead by ethylenediaminetetraacetate." *J. Pharmacol. Exp. Ther.,* 157:196–206, 1967. (17)

Hammond, P. B., Aronson, A. L., and Olson, W. C. "The relationship between inhibition of omega-aminolevulinic acid dehydratase by lead and lead mobilization by ethylenediamine tetraacetate (EDTA)." *Toxicol. Appl. Pharmacol.,* 26:466–75, 1973. (18)

Hansotia, P., Peters, H., Bennett, M., and Brown, R. "Chelation therapy in Wegener's granulomatosis. Treatment with EDTA." *Ann. Otol.,* 78:388–402, 1969. (27)

Hardcastle, P. T., and Eggenton, J. "The effect of EDTA on the electrical activity of rat jejunum." *Biochem. Biophys. Acta.,* 241:930–33, 1971. (14)

Hardy, H. L. "Clinical experience with the use of calcium disodium ethylenediaminetetraacetate in the therapy of lead poisoning." *Fed. Proc.*, 20 (Suppl. 10): 199–205, September 1961. (11)

Harman, J. W., and MacBrinn, M. C. "The effect of phenazine methosulphate, pyocyanine and EDTA on mitochondrial succinic dehydrogenase." *Biochem. Pharmacol.*, 12:1265–78, 1963. (22)

Harris, C. C., and Leone, C. A. "Some effects of EDTA and tetraphenylboron on the ultrastructure of mitochondria in mouse liver cells." *J. Cell. Biol.*, 28:405–8, 1966. (11)

Hastings, A. B. "Introductory remarks. [Proceedings of a conference on biological aspects of metal-binding.]" *Fed. Proc.*, 20 (Suppl. 10): 1–3, September 1961. (0)

Haumont, S., and Vincent, J. "Action of calcium versenate on lead fixed in vivo in compact bone." *Exp. Cell. Res.*, 18:404–406, 1959. (6)

Hausmann, E. "Change in plasma phosphate concentration on infusion of calcium gluconate or Na_2-EDTA." *Proc. Soc. Exp. Biol. Med.*, 134:182–84, 1970. (12)

Hawkins, W. W., Leonard, V. G., Maxwell, J. E., and Rastogi, K. S. "A study of the prolonged intake of small amounts of ethylenediaminetetraacetate on the utilization of low dietary levels of calcium and iron by the rat." *Canad. J. Biochem.*, 40:391–95, 1962. (5)

Heath, O. V. S., and Clark, J. E. "Chelating agents and auxin." *Nature*, 201:585–87, 1964. (9)

Heling, B, Shapiro, S., and Sciaky, I. "An in vitro comparison of the amount of calcium removed by the disodium salt of EDTA and hydrochloric acid during endodontic procedures." *Oral surg.*, 19:531–33, 1965. (6)

Heller, J., and Vostal, J. "Renal excretion of calcium-disodium-ethylenediaminetetraacetic acid—a new tubular secretory mechanism?" *Experientia.*, 2099–101, 1964.(8)

Hemmens W. F. "Effects of ethylenediamine tetraacetic acid on the cell fragility of brewer's yeast." *Nature*, 200:383–84, 1963. (3)

Henriques, O. B. "Effect of 1,10-phenantroline and EDTA on bradykinin assay on the guinea-pig ileum." *Biochem. Pharmacol.*, 20:2759–63, 1971. (5)

Herd, J. K., and Vaughan, J. H. "Calcinosis universalis complicating dermatomyositis—its treatment with Na_2 EDTA. Report of two cases in children." *Arthritis Rheum.*, 7:259–71, 1964. (53)

Hilfer, S. R., and Hilfer, E. K. "Effects of dissociating agents on the fine structure of embryonic chick thyroid cells." *J. Morph.*, 119:217–31, 1966. (23)

Hodgkinson, R. "A comparative study of iron absorption and utilization following ferrous sulphate and sodium ironedetate ('Sytron')." *Med. J. Aust.*, 1:809–11, 1961. (6)

Hollister, L. E., Cull, V. L., Gonda, V. A., and Kolb, F. O. "Hepatolenticular degeneration. Clinical, biochemical and pathologic study of a patient with fulminant course aggravated by treatment with BAL and versenate." *Am. J. Med.*, 28:623–30, 1960. (16)

Hori, S. H. "Effect of EDTA on histochemical demonstration of phosphorylase activity." *J. Histochem. Cytohem.*, 14:501–8, 1966. (32)

Hosli, P. "Therapy of scleroderma with the disodium salt of ethylenediaminetetraacetic acid; a contribution to the toxicology of versenate. Part I." *Arzeimittelforschung*, 10:65–74, 1960. (0)

Hovig, T., Nicolaysen, A. and Nicolaysen, G. "Ultrastructural studies of the alveolar-capillary barrier in isolated plasma-perfused rabbit lungs. Effects of

EDTA and of increased capillary pressure." *Acta. Physiol. Scand.*, 82:417–31, 1971. (13)

Huddart, H. "Superprecipitation of actomyosin extracted from crayfish skeletal muscle: the effect of heavy metal cations, EDTA and drugs." *J. Exp. Zool.*, 177:407–15, 1971. (24)

Hultin, T., and Ostner, U. "Specific unmasking of ribosomal proteins under the influence of chelating agents and increased ionic strength." *Biochem. Biophys. Acta.*, 160:229–38, 1968. (19)

Hurych, J., and Nordwig, A. "Inhibition of collagen hydroxylysine formation by chelating agents." *Biochem. Biophys. Acta.*, 140:168–70, 1967. (11)

Itoh, H., Yamaguchi, T., and Yamasawa, S. "The inhibition of hematopoietic action of iron by ethylenediamine tetraacetic acid (EDTA)." *Yokohama Med. Bull.*, 13:9–16, 1962. (6)

Jackson, S. H., and Brown, F. "Simultaneous determination of calcium and magnesium of serum by a single chelometric titration." *Clin. Chem.*, 10:159–69, 1964. (11)

Jick, S., and Karsh, R. "The effect of calcium chelation on cardiac arrhythmias and conduction disturbances." *Am. J. Cardiol.*, 4:287–93, 1959. (18)

Johnson, L. A., and Steven, M. J., Ed. "Proceedings of a conference on biological aspects of metal-binding, held at the Pennsylvania State University, Unversity Park, Pennsylvania, September 6-9, 1960." *Fed. Proc.*, 20 (Suppl. 10): i–ix, 1–273, September 1961.

Johnson, S. A. M. "Use of chelating agent edathamil disodium in acrosclerosis, sarcoidosis and other skin conditions with comments on tryptophan metabolism in sarcoidosis." *Wisconsin Med. J.*, 59:651–55, 1960 (0)

Johnstone, M. A., Sullivan, W. R., and Grant, W. M. "Experimental zinc chloride ocular injury and treatment with disodium edetate." *Am. J. Ophthal.*, 76:137–42, 1973. (3)

Jones, K. H., and Fourman, P. "Edetic-acid test of parathyroid insufficiency." *Lancet*, 2:119–21, 1963. (7)

Kahn, V., and Blum, J. J. "The glycogen phosphorylase of *Tetra + hymena pyriformis*. II. "Inhibition and inactivation by EDTA and ATP and other kinetic properties." *Arch. Biochem. Biophys.*, 143:92–105, 1971. (35)

Kalliomaki, J. L., Markkanen, T. K., and Mustonen, V. A. "Serum calcium and phosphorus homeostatis in man studied by means of the sodium-EDTA test." *Acta. Med. Scand.*, 170:211–14, 1961. (9)

Kaminski, E. E., and Pacenti, D. M. "Spectrophotometric titration of edetic acid in ophthalmic solutions." *J. Pharm. Sci.*, 63:1133–36, 1974. (8)

Karasawa, T., Funakoshi, H., Jurukawa, K., and Yoshida, K. "EDTA prevents the photo-catalyzed destruction of the products of catecholamine oxidation." *Anal. Biochem.*, 53:278–81, 1973. (11)

Kay, J. E. "Interaction of lymphocytes and phytohaemagglutinin: inhibition by chelating agents." *Exp. Cell. Res.*, 68:11–16, 1971. (25)

Keatinge, W. R. "Responsiveness of arterial smooth muscle to noradrenaline after long periods in Ca-free solution containing EDTA at low temperature." *J. Physiol.*, 216:31P, 1971. (3)

Kebe, S. R. "ACTH and a chelating agent for schizophrenia." *West. Med.*, 4:46, 48, 1963. (6)

Keech, M. K., McCann, D. S., Boyle, A. J., and Pinkus, H. "Effect of ethylenediaminetetraacetic acid (EDTA) and tetrahydroxyquinone on sclerodermatous skin. Histologic and chemical studies." *J. Invest. Derm,* 47:235–46, 1966. (66)

Kehoe, R. A. "Value of calcium disodium ethylenediaminetetraacetate and British anti-lewisite in therapy of lead poisoning." *Fed. Proc.,* 20 (Suppl. 10): 196–99, September 1961. (5)

Kelly, H. G., Turton, M. R., and Hatcher, J. D. "Renal and cardiovascular effects induced by intravenous infusion of magnesium chelates." *Canad. Med. Assoc. J.,* 84:1124–28, 1961. (6)

Kemble, J. V. H. "The new chelating agent Ca-DTPA in the treatment of primary haemochromatosis." *Guy Hosp. Rep.,* 113:68–73, 1964. (24)

Kerby, G. P., and Taylor, S. M. "The role of chelation and of human plasma in the uptake of serotonin by human platelets." *J. Clin. Invest.,* 40:44–51, 1961. (15)

King, L. R., Portnoy, R. M., Goldsmith, R. E., and Zalme, E. "Serum calcium homeostatis following thyroid surgery as measured by ethylenediamine tetra-acetate infusion." *J. Clin. Endocr.,* 25:577–84, 1965. (13)

Kirschbaum, J., and Wainio, W. W. "Release of sulfhydryl groups in cytochrome oxidase by ethylenediaminetetraacetic acid." *Biochem. Biophys. Acta.,* 118:643–44, 1966. (9)

Kissmeyer-Nielsen, F., and Andresen, E. "EDTA-Na_2 and leukocytes." *Bibl. Haemat.,* 19:434–38, 1964. (1)

Kitchell, J. R., Palmon, Jr., F., Aytan, N., and Meltzer, L. E. "The treatment of coronary artery disease with disodium EDTA. A reappraisal." *Am. J. Cardiol.,* 11:501–6, 1963. (7)

Kneller, L. A., Uhl, H. S. M., and Brem, J. "Successful calcium disodium ethylene diamine tetra-acetate treatment of lead poisoning in an infant." *New Eng. J. Med.,* 252:338–40, 1955. (16)

Koen, A., McCann, D. S. and Boyle, A. J. "Some in vivo effects of chelation. II. Animal experimentation." *J. Chron. Dis.,* 16:329–33, 1963. (8)

Koike, T. I., Kratzer, F. H., and Vohra, P. "Intestinal absorption of zinc or calcium-ethylenediaminetetraacetic acid complexes in chickens." *Proc. Soc. Exp. Biol. Med.,* 117:483–86, 1964. (15)

Krahl, M. E. "Insulinlike and anti-insulin effects of chelating agents on adipose tissue." *Fed. Proc.,* 25:832–34, 1966. (9)

Kuo, J. F. "Differential affects of Ca^{2+}, EDTA and adrenergic blocking agents on the actions of some hormones on adenosine 3′,5′-monophosphate levels in isolated adipose cells as determined by prior labeling with [8–^{14}C]-adenine." *Biochem. Biophys. Acta.,* 208:509–16, 1970. (26)

Kurihara, M., and Sano, S. "Reduction of cytochrome c by ferrous ions and ethylenediaminetetraacetic acid in acid solution." *J. Biol. Chem.,* 245:4804–06, 1970. (8)

Lacombe, M. L. and Hanoune, J. "Enhanced specificity of epinephrine binding by rat liver plasma membranes in the presence of EDTA." *Biochem. Biophys. Res. Commun.,* 58:667–73, 1974. (15)

Lahaye, D., Roosels, D., and Verwilghen, R. "Diagnostic sodium calcium-edetate mobilization test in ambulant patients." *Brit. J. Indust. Med.,* 25:148–49, 1968. (5)

Lamar, C. P. "Chelation endarterectomy for occlusive atherosclerosis." *J. Am. Geriat. Soc.,* 14:272–94, 1966. (28)

Lamar, C. P. "Chelation therapy of occlusive arteriosclerosis in diabetic patients." *Angiology,* 15:379–95, 1964. (33)

Lampasso, J. A. "Changes in hematologic values induced by storage of edhylenediaminetetraacetate human blood for varying periods of time.:" *Am. J. Clin. Path.*, 49:443–47, 1968. (11)

Larsen, B . A., Bidwell, R. G., and Hawkins, W. W. "The effect of ingestion of disodium ethylenediaminetetraacetate on the absorption and metabolism of radioactive iron by the rat." *Canad. J. Biochem.*, 38:51–55, 1960. (11)

Larsen, B. A., Hawkins, W. W., Leonard, V. G., and Armstrong, J. E. "The effect of prolonged intake of ethylenediaminetetraacetate on the utilization of calcium and iron by the rat." *Canad. J. Biochem.*, 38:813–17, 1960. (7)

Larson, R. H. "The effect of EDTA on the pattern of caries development and its association with biologic changes in the rat." *J. Dent. Res.*, 38:1207–12, 1959. (16)

Larson, R. H., Zipkin, I., and Rubin, M. "Effect of administration of EDTA by various routes on dental caries in the rat.. Possible role of coprophagy." *Arch. Oral Biol.*, 5:49–54, 1961. (22)

Laszt, L. "Influence of ethylenediamine tetraacetate and calcium ions on vascular tension." *Nature*, 212:1587, 1966. (18)

Lavender, A. R., Pullman, T. N., and Goldman, D. "Spectrophotometric determination of ethylenediaminetetraacetic acid in plasma and urine." *J. Lab. Clin. Med.*, 63:299–305, 1964. (8)

Leckert, J. T., McHardy, G. G., and McHardy, R. J. "Edathamil (EDTA) therapy of interstitial calcinosis." *South. Med. J.*, 53:728–31, 1960. (10)

Ledbetter, J. R., and Enneking, W. F. "Effects of sodium ethylenediamine tetraacetic acid on fracture healing." *Surg. Forum.*, 15:446–48, 1964. (0)

Lee, C. P. "A fluorescent probe of the hydrogen ion concentration in ethylenediaminetetraacetic acid particles of beef heart mitochondria." *Biochemistry*, 10:4375–81, 1971. (39)

Lee, C. P., Ernster, L., and Chance, B. "Studies of the energy-transfer system of submitochondrial particles. Kinetic studies of the effect of oligomycin on the respiratory chain of EDTA particles." *Europ. J. Biochem.*, 8:153–63, 1969. (25)

Leive, L. "A nonspecific increase in permeability in *Escherichi coli* produced by EDTA." *Proc. Natl. Acad. Sci. USA*, 53:745–50, 1965. (18)

Leive, L. "Release of lipopolysaccharide by EDTA treatment of E. coli." *Biochem. Biophys. Res. Commun.*, 21:290–96, 1965. (17)

Leive, L. "Studies on the permeability change produced in coliform bacteria by ethylenediaminetetraacetate." *J. Biol. Chem.*, 243:2373–80, 1968. (46)

Leive, L., Shovlin, V. K., and Mergenhagen, S. E. "Physical, chemical, and immunological properties of lipopolysaccharide released from *Escherichia coli* by ethylenediaminetetraacetate." *J. Biol. Chem.* 243:6384–91, 1968. (32)

Lenta, P., and Riehl, A. "The influence of metallic chelates on the diphosphopyridine nucleotide oxidase and diphosphopyridine nucleotide-cytochrome c reductase systems." *J. Biol. Chem.*, 235:859–64, 1960. (24)

Lepke, S., and Passow, H. "The inhibition of potassium loss of fluoride-poisoned erythrocytes by chelating agents." *Pflueger Arch. Ges. Physiol.*, 271:389–96, 1960. (17)

Liberman, U. A., Barzel, U., De Vries, A., and Ellis, H. "Myositis ossificans traumatica with unusual course. Effect of EDTA on calcium, phosphorus and manganese escretion." *Am J., Med. Sci.*, 254:35–47, 1967. (50)

Lochte, Jr., H. L., Ferrebee, J. W., and Thomas, E. D. "The effect of heparin and

EDTA on DNA synthesis by marrow in vitro." *J. Lab. Clin. Med.*, 55:435–38, 1960. (11)

Lockefeer, J. H., Hackeng, W. H. L., and Birdenhagër, J. C. "Parathyroid hormone secretion in disorders of calcium metabolism studied by means of EDTA." *Acta. Endocr.*, 75:286–96, 1974. (18)

Lynch, K. L., and Moskowitz, M. "Effects of chelates in chemotherapy of experimental gas-gangrene toxemia." *J. Bact.*, 96:1925–30, 1968. (13)

Lynn, Jr., W. S., Fortney, S., and Brown, R. H. "Role of EDTA and metals in mitochondrial contraction." *J. Cell. Biol.*, 23:9–19, 1964. (26)

McBurney, L. J., and Radomski, M. W. "The effects of washing, EDTA, magnesium and calcium on oxidate phosphorylation and respiratory rates of mitochondria from heat- and cold-acclimated rats." *Comp. Biochem. Physiol.*, 44B:1219–33, 1973. (59)

McCann, D. S., Koen, Z., Zdybek, G., and Boyle, A. J. "Effect of chelation on phosphorus metabolism in experimental atherosclerosis." *Circ. Res.*, 11:880–84, 1962. (7)

McCoy, J. E., Carre, I. J., and Freeman, M. "A controlled trial of edathamil calcium disodium in acrodynia." *Pediatrics*, 25:304–8, 1960. (24)

McLaren, J. R., Galambos, J. T., and Drew, W. D. "Hepatic and renal studies with iron-50 EDTA in patients with and without liver or kidney disease." *Radiology*, 81:447–54, 1963. (7)

Marcelle, R., and Lecomte, J. "On the cardiovascular activities of the sodium salt of ethylenediaminetetraacetic acid." *C. R. Soc. Biol.*, 153:1483–85, 1959. (1)

Marlow, C. G., and Sheppard, G. "[^{51}Cr EDTA,][hydroxymethyl-^{14}C] inulin and inulin-T for the determination of glomerular filtration rate." *Clin. Chem. Acta.*, 28:479–88, 1970. (10)

Marney, Jr., S. R., and Des Prez, R. M. "Rabbit platelet injury by soluble antigen and antibody. III. Effects of the sodium and magnesium salts of EDTA and EGTA." *J. Immun.*, 106:1447–52, 1971. (16)

Marsh, L., and Fraser, F. C. "Chelating agents and teratogenesis." (Letters) *Lancet*, 1:846, 1973. (29)

Martell, A. E. "Opening remarks by the chairman. [Proceedings of a conference on biological aspects of meta-binding]." *Fed. Proc.*, 20 (Suppl. 10): 4, September 1961. (0)

Martell, A. E. "Some factors governing chelating tendencies and selectivities in their interaction of ligands with metal ions." *Fed. Proc.*, 20 (Suppl. 10): 35–39, September 1961. (9)

Martin, R. B. "Metal ion binding to peptides and proteins." *Fed. Proc.*, 20 (Suppl 10): 54–59, September 1961. (11)

Maruyama, K., and Gergely, J. "Removal of the bound calcium of G-actin by ethylenediaminetetraacetate (EDTA)." *Biochem. Biophys. Res. Commun.*, 6:245–49, 1961. (11)

Maxwell, G. M., Elliott, R. B., and Robertson, E. "The effect of Na_3EDTA-induced hypocalcemia upon the general and coronary hemodynamics of the intact animal." *Am. Heart J.*, 66:82–87, 1963. (13)

Meltzer, L. E., Kitchell, J. R., and Palmon, Jr., F. "The long term use, side effects, and toxicity of disodium ethylenediamine-tetraacetic acid (EDTA)." *Am. J. Med Sci.* 242:11–17, 1961. (37)

Meltzer, L. E., Palmon, Jr., F. P., and Kitchell, J. R. "Hypoglycaemia induced by disodium ethylenediamine tetra-acetic acid." *Lancet*, 2:637–38, 1961. (4)

Michaels, G. B., and Eagon, R. G. "The effect of ethylenediaminetetraacetate and of lysozyme on isolated lipopolysaccharide from *Pseudomonas aeruginosa*." *Proc. Soc. Exp. Biol. Med.*, 122:866–68, 1966. (12)

Moqtaderi, F., Himal, H. S., Rudick, J., and Dreiling, D. A. "Pancreatic transductal electrolyte flux. Effect of EDTA." *Am. J. Gastroent.*, 58:177–84, 1972. (16)

Morasca, L., Balconi, G., Dolfini, E., and Oldani, C. "Effect of trypsin and ethylenediamine tetra acetic acid on the rate of adhesion of KB cells." *Experientia*, 29:1005–1006, 1973. (9)

Morgan, J. M. "Chelation therapy in lead nephropathy." *South Med. J.*, 68:1001–1006, 1975. (18)

Myers, H. L. "Topical chelation therapy for varicose pigmentation." *Angiology*, 17:66–68, 1966. (8)

Nakano, J., Cole, B., and Ishii, T. "Effects of disodium EDTA on the cardiovascular responses to prostaglandin E_1." *Experientia*, 24:808–809, 1968. (5)

Nayler, W. G. "Ventricular arrhythmias following the administration of Na_2EDTA." *J. Pharmacol. Exp. Ther.*, 137:5–13, 1962.

Nedergaard, O. A., and Vagne, A. "Effect of EDTA and other chelating agents on norepinephrine uptake by rabbit aorta in vitro." *Proc. West. Pharmacol. Soc.*, 11:87–90, 1968. (3)

Neldner, K. H., Winkelmann, R. K., and Perry, H. O. "Scleroderma. An evaluation of treatment with disodium edetate." *Arch. Derm.*, 86:305–09, 1962. (14)

Neu, H. C. "The role of amine buffers in EDTA toxicity and their effect on osmotic shock." *J. Gen. Microbiol.*, 57:215–20, 1969. (11)

Offer, G. W. "The antagonistic action of magnesium ions and ethylenediamine tetraacetate on myosin A ATPase (potassium activated)." *Biochem. Biophys. Acta.*, 89:566–69, 1964. (25)

Okuda, K., and Sasayama, K. "Effects of ethylenediaminetetraacetate and metal ions in intestinal absorption of vitamin B_{12} in man and rats." *Proc. Soc. Exp. Biol. Med.*, 120:17–20, 1965.

Olwin, J. H., and Koppel, J. L. "Reduction of elevated plasma lipid levels in atherosclerosis following EDTA therapy." *Proc. Soc. Exp. Biol. Med.*, 128:1137–40, 1968. (5)

Ontka, J. A. "Physical and chemical changes in isolated chylomicrons: prevention by EDTA." *J. Lipid. Res.*, 11:367–75, 1970. (32)

Oser, B. L. "Observations on the chronic ingestion of a chelating agent by rats and dogs." *Fed. Proc.*, 20 (Suppl. 10): 158, September 1961. (0)

Oser, B. L., Oser, M., and Spencer, H. C. "Safety evaluation studies of calcium EDTA." *Toxicol. Appl. Pharmacol.*, 5:142–62, 1963. (4)

O'Sullivan, W. J., and Morrison, J. F. "The effect of trace metal contaminants and EDTA on the velocity of enzyme-catalysed reactions. Studies on ATP: Creatine phosphotransferase." *Biochem. Biophys. Acta.*, 77:142–44, 1963. (5)

Painter, J. T., and Morros, E. J. "Porphyria. Its manifestations and treatment with chelating agents." *Texas State J. Med.*, 55:811–18, 1959. (15)

Parfitt, A. M. "Study of parathyroid function in man by EDTA." *J. Clin. Endocr.*, 29:569–80, 1969. (42)

Pavek, K., Drimal, J., and Selecky, F. V. "Circulatory effects of disodium edetate in digoxin-induced ventricular tachycardia." *Cardiologia*, 50:297–304, 1967. (13)

Payne, J. M., and Sansom, B. F. "The relative toxicity in rats of disodium ethylene diamine tetra-acetate, sodium oxalate and sodium citrate." *J. Physiol.*, 170:613–20, 1964. (2)

Perry, Jr., H. M. "Chelation therapy in circulatory and sclerosing diseases." *Fed. Proc.*, 20 (Suppl. 10): 254–57, September 1961.

Perry, Jr., H. M., and Schroeder, H. A. "Depression of cholesterol levels in human plasma following ethylenediamine tetraacetate and hydralazine." *J. Chron. Dis.*, 2:520–33, 1955. (20)

Perry, Jr., H. M., and Schroeder, H. A. "Lesions resembling vitamin B complex deficiency and urinary loss of zinc produced by ethylenediamine tetraacetate." *Am. J. Med.*, 22:168–72, 1957. (11)

Perry, S. V., Davies, V., and Hayter, D. " 'Natural' tropomyosin and the factor sensitizing actomyosin adenosine triphophatase to ethylenedioxybis (ethyleneamino) tetraacetic acid." *Biochem. J.*, 99:1C–2C, 1966. (14)

Peters, H. A. "Porphyric psychosis and chelation therapy." *Recent Adv. Biol. Psychiat.*, 4:204–17, 1961. (31)

Peters, H. A. "Trace minerals, chelating agents and the porphyrias." *Fed. Proc.*, 20 (Suppl. 10): 227–34, September 1961. (42)

Peters, H. A., Johnson, S. A. M., Cam, S., Oral, S., Muftu, Y., and Ergene, T. "Hexachlorobenzene-induced porphyria: effect of chelation on the disease, porphyrin and metal metabolism." *Am. J. Med. Sci.*, 251:314–422, 1966. (44)

Philippot, J., and Authier, M. H. "Study of human red blood cell membrane using sodium deoxycholate. II. Effects of cold storage, EDTA and small deoxycholate concentrations on ATPase activities." *Biochem. Biophys. Acta.*, 298:887–900, 1973. (32)

Pho, D. B., and Bethune, J. L. "The effect of 2-mercapto-ethanol (ME) and EDTA on the sub-species structure of horse liver alcohol dehydrogenase (LADH) in 8 M urea." *Biochem. Biophys. Res. Commun.*, 47:419–25, 1972. (28)

Porte, J., Brard, E., Roche, M., and Santini, R. "Effects of prolonged administration of disodic tetracemate and thyrocalcitonin on the parathyroid gland." *C. R. Soc. Biol.*, 165:100–103, 1971. (2)

Powell, G. W., Miller, W. J., and Blackmon, D. M. "Effects of dietary EDTA and cadmium on absorption, excretion and retention of orally administered ^{65}Zn in various tissues of zinc-deficient and normal goats and calves." *J. Nutr.*, 93:203–12, 1967. (34)

Price, J. M. "Some effects of chelating agents on tryptophan metabolism in man." *Fed. Proc.*, 20 (Suppl. 10):223–26, September 1961. (16)

Ragan, H. A. "Platelet agglutination induced by ethylenediaminetetraacetic acid in blood samples from a miniature pig." *Am J. Vet. Res.*, 33:2601–2603, 1972. (4)

Rahman, Y. E., Rosenthal, M. W., and Cerny, E. A. "Intracellular plutonium: removal by liposome-encapsulated chelating agent." *Science*, 180:300–302, 1973. (20)

Rahman, Y. E., Rosenthal, M. W., Cerny, E. A., and Moretti, E. S. "Preparation and prolonged tissue retention of liposome-encapsulated chelating agents." *J. Lab. Clin. Med.*, 83:640–47, 1974. (22)

Rajabalee, F. J. M., Potvin, M., and Laham, S. "Separation of NTA and EDTA chelates by thin-layer chromatography." *J. Chromatogr.*, 79:375–79, 1973. (10)

Ramberg, Jr., C. F., Mayer, G. P., Kronfeld, D. S., Aurback, G. D., Sherwood, L.

M., and Potts, Jr., J. T. "Plasma calcium and parathyroid hormone responses to EDTA infusion in the cow." *Am. J. Physiol.*, 213:878–82, 1967. (21)

Ravnik, C., Sand, H. F., and Morch, T. "Enamel lesions produced in vitro by solutions of EDTA and EDTA-sodium salts." *Acta. Odont. Scand.*, 20:349–58, 1962. (6)

Raymond, J. Z., and Gross, P. R. "EDTA: preservative dermatitis." *Arch. Derm.*, 100:436–40, 1969. (12)

Reber, K., and Studer, A. "Acceleration of the gastrointestinal absorption of heparin by calcium-binding substances." *Experientia*, 19:141–42, 1963. (5)

Reger, J. F. "The fine structure of the extensor digitorum longus IV muscle of *Rana pipiens* following calcium removal via EDTA." *Exp. Cell. Res.*, 43:435–43, 1966. (21)

Renoux, M., and Mikol, C. "Chelation." *Presse Med.*, 72:3117–19, 1964. (0)

Reuber, M. D. "Accentuation of Ca edetate nephrosis by cortisone." *Arch. Path.*, 76:382–86, 1963. (18)

Reuber, M. D. "Calcium disodium edetate nephrosis in female rats of varying ages." *J. Path.*, 97:335–38, 1969. (16)

Reuber, M. D. "Hepatic lesions in young rats given calcium disodium edetate." *Toxicol. Appl. Pharmacol.*, 11:321–26, 1967. (14)

Reuber, M. D., and Bradley, J. E. "Acute versenate nephrosis occurring as the result of treatment for lead intoxication." *JAMA*, 174:263–69, 1960. (17)

Reuber, M. D., and Lee, C. W. "Calcium disodium edetate nephrosis in inbred rats. Variation in Marshall, Buffalo, Fischer, and ACI strains." *Arch. Environ. Health*, 13:554–57, 1966.

Reuber, M. D., and Schmieler, G. C. "Edetate kidney lesions in rats." *Arch. Environ. Health*, 5:430–36, 1962. (23)

Reuter, H., Niemeyer, G., and Gross, R. "The distribution of EDTA and citrate in blood and plasma." *Throm. Diath. Haemorrh.*, 19:213–20, 1968. (3)

Rice, E. G., Greenberg, S. M., Herndon, J. F., and VanLoon, E. J. "Effect of ethylenediamine tetra-acetate on vitamin B_{12} absorption in the rat." *Nature*, 184:1948, 1959. (5)

Rinard, G. A. "Phosphorylase activity in rat uterine homogenates. Loss of activity related to in vivo treatment with estrogen, progesterone, relaxin and CaEDTA." *Biochem. Biophys. Acta*, 222:455–64, 1970. (20)

Robotti, C., and Lovisolo, D. "Pyrophosphate and ethylenediaminetetraacetate as relaxants for lower invertebrates prior to fixation." *Stain Techn.*, 47:37–38, 1972. (6)

Rosenbaum, J. L., Mason, D., and Seven, M. J. "The effect of disodium EDTA on digitalis intoxication." *Am. J. Med. Sci.*, 240:77–84, 1960. (14)

Rosenberg, J. S., Tashima, Y., and Horecker, B. L. "Activation of rabbit kidney fructose diphosphatase by Mg-EDTA, Mc-EDTA and Co-EDTA complexes." *Arch. Biochem. Biophys.*, 154:283–91, 1973. (30)

Rosoff, B., Hart, H., Methfessel, A. H., and Spencer, H. "Fate of intravenously administered zinc chelates in man." *J. Appl. Physiol.*, 30:12–16, 1971. (30)

Rossi, E. C. "Effects of ethylenediaminetetraacetate (EDTA) and citrate upon platelet glycolysis." *J. Lab. Clin. Med.*, 69:204–16, 1967. (24)

Rubin, H. "Inhibition of DNA synthesis in animal cells by ethylene diamine tetraacetate, and its reversal by zinc." *Proc. Natl. Acad. Sci. USA*, 69:712–16, 1972. (15)

Rubin, M. "Design of chelates for therapeutic objectives." *Fed. Proc.*, 20 (Suppl. 10): 149–57, September 1961. (54)

Russell, A. D. "Effect of magnesium ions and ethylenediamine tetra-acetic acid on the activity of vancomycin against *Escherichia coli and Staphylococcus aureus.*." *J. Appl. Bact.*, 30:395–401, 1967. (17)

Saito, S., Bush, I. M., Mackenzie, A. R., and Whitmore, Jr., W. F. "The effects of certain metals and chelating agents on the motility of dog epididymal and ejaculated spermatozoa." *Invest. Urol.*, 4:546–55, 1967. (28)

Sand, H. F. "The dissociation of EDTA and EDTA-sodium salts." *Acta. Odont. Scand.*, 19:469–82, 1961. (6)

Sanders, Jr., S. M. "Plutonium excretion. Study following treatment with zirconium citrate and edathamil calcium-disodium." *Arch. Environ. Health*, 2:474–83, 1961. (27)

Sanui, H., and Pace, N. "Effect of ATP, EDTA and EGTA on the simultaneous binding of Na, K, Mg and Ca by rat liver microsomes." *J. Cell. Physiol.*, 69:11–19, 1967. (22)

Saumur, J. "Inhibition of L. E. phenomenon by EDTA." *J-Lancet*, 82:240–42, 1962. (9)

Saunders, J. F., Princiotto, J. V., and Rubin, M. "Effect of calcium disodium ethylenediamine tetraacetate on hypercholesterolemic rabbits." *Proc. Soc. Exp. Biol. Med.*, 92:29–31, 1956. (9)

Sawai, Y., Makino, M., Miyasaki, S., Kawamura, Y., Mitsuhashi, S., and Okonogi, T. "Studies on the improvement of treatment of Habu snake (*Trimeresurus flavoviridis*) bite. 2. Antitoxic action of monocalcium disodium ethylene diamine tetraacetate on Habu venom." *Jap. J. Exp. Med.*, 31:267–75, 1961. (8)

Sawyer, D. T. "Infrared spectra and correlations for the ethylenediaminetetraacetic acid metal chelates." *Ann. NY Acad. Sci.*, 88:302–21, 1960. (24)

Schane, H. P. "Estrogen-like effect of chelating agents on rat uterine phosphorylase activity." *Endocrinology*, 76:491–98, 1965. (18)

Schanker, L. S., and Johnson, J. M. "Increased intestinal absorption of foreign organic compounds in the presence of ethylenediaminetetraacetic acid (EDTA)." *Biochem. Pharmacol.*, 8:421–22, 1961. (4)

Schroeder, H. A., Nason, A. P., and Mitchener, M. "Action of a chelate of zinc on trace metals in hypertensive rats." *Am. J. Physiol.*, 214:796–800, 1968. (13)

Schubert, J. "Chelation in medicine." *Sci. Am.*, 214:40–50, 1966. (0)

Schubert, J. "Radioelement removal by chelating agents: application of mass action laws and other factors." *Fed. Proc.*, 20 (Suppl. 10): 219–22, September 1961. (5)

Schwartz, S. L., and Bond, J. C. "Induction of vacuolation in the mouse peritoneal macrophage by EDTA." *Exp. Cell. Res.*, 66:253–57, 1971. (7)

Schwartz, S. L., Hayes, J. R., Ide, R. S., Johnson, C. B., and Doolan, P. D. "Studies of the nephrotoxicity of ethylenediaminetetraacetic acid." *Biochem. Pharmacol.*, 15:377–89, 1966. (24)

Schwartz, S. L., Johnson, C. B., and Doolan, P. D. "Study of the mechanism of renal vacuologenesis induced in the rat by ethylenediaminetetraacetate. Comparison of the cellular activities of calcium and chromium chelates." *Mol. Pharmacol.*, 6:54–60, 1970. (11)

Schwartz, S. L., Johnson, C. B., Hayes, J. R., and Doolan, P. D. "Subcellular localization of ethylenediaminetetraacetate in the proximal tubular cell of the rat kidney." *Biochem. Pharmacol.*, 16:2413–19, 1967. (17)

Seidel, J. C., and Gergely, J. "Studies on myofibrillar adenosine triphosphatase with calcium-free adenosine triphosphate. I. The effect of ethylenediaminetetraacetate, calcium, magnesium, and adenosine triphosphate." *J. Biol. Chem.*, 238:3648–53, 1963. (34)

Selander, S. "Treatment of lead poisoning. A comparison between the effects of sodium calcium-edetate and penicillamine administered orally and intravenously." *Brit. J. Indust. Med.*, 24:272–82, 1967. (35)

Settlemire, C. T., Hunter, G. R., and Brierley, G. P. "Ion transport in heart mitochondria. XIII. The effect of ethylenediaminetetraacetate on monovalent ion uptake." *Biochem. Biophys. Acta.*, 162:487–99, 1968. (27)

Shapiro, S., Perez, G., Gedalia, I., and Sulman, F. G. "The solubility rate of calcified dental tissues and of calculus in sodium EDTA solutions." *J. Periodont.*, 39:9–10, 1968. (10)

Sheikh, M. A., and Parker, M. S. "The influence of EDTA and PEA upon the fungistatic action of aminacrine hydrochloride." *J. Pharm. Pharmacol.*, 24:158P, 1972. (2)

Sheldon, J., and Taylor, K. W. "The immunoassay of insulin in human serum treated with sodium ethylene diamine tetraacetate." *J. Endocr.*, 33:157–58, 1965. (5)

Sheppard, B. L. "Platelet adhesion in the rabbit abdominal aorta following the removal of endothelium with EDTA." *Proc. Roy. Soc. Lond.*, (Biol) 182:103–108, 1972. (32)

Shibata, S., and Kobayashi, B. "Calcium ion-induced swelling of EDTA-treated, washed blood platelets of rats." *Life Sci.*, 8:727–32, 1969. (8)

Shiloah, J., Gedalia, I., Shapiro, S., and Jacobowitz, B. "Solubility of calcified dental tissues and of calculus in EDTA and 2%NaF." *J. Dent. Res.*, 52:845, 1973. (7)

Shrand, H. "Treatment of lead poisoning with intramuscular edathamil calcium-disodium." *Lancet*, 1:310–12, 1961. (25)

Simmelink, J. W., Nygaard, V. K., and Scott, D. B. "Theory for the sequence of human and rat enamel dissolution by acid and by EDTA: a correlated scanning and transmission electron microscope study." *Arch. Oral Biol.*, 19:183–97, 1974. (30)

Simpson, K., and Blunt, A. "Acute ferrous sulphate poisoning treated with edathamil calcium-disodium." *Lancet*, 2:1120–22, 1960. (15)

Sincock, A. M. "Life extension in the rotifer *Mytilina brevispina var redunca* by the application of chelating agents." *J. Geront.* 30:289–93, 1975. (5)

Sivjakov, K. I., and Braun, H. A. "The treatment of acute selenium, cadmium, and tungsten intoxication in rats with calcium disodium ethylenediamine-tetraacetate." *Toxicol. Appl. Pharmacol.*, 1:602–608, 1959. (15)

Sjogvist, A., and Hultin, T. "Conformational effects of mercurials on rat liver ribosomes: a comparison between the unmasking of a shielded protein in the 60S subunit by phenyl mercurials and EDTA." *Chem-Biol. Interact.*, 6:131–48, 1973. (36)

Smith, C. L. "The action of EDTA on the succinoxidase systems of liver mitochondria from lower vertebrates." *Comp. Biochem. Physiol.*, 25:805–20, 1968. (25)

Smith, H. D., King, L. P., and Margolin, E. G. "Treatment of lead encephalopathy. The combined use of edetate and hemodialysis." *Am. J. Dis. Child.*, 109:322–24, 1965. (6)

Snow, C., and Allen, A. "The release of radioactive nucleic acids and mucoproteins by trypsin and ethylenediaminetetra-acetate treatment of baby-hamster cells in the tissue culture." *Biochem. J.*, 119:707–14, 1970. (31)

Soffer, A., and Toribara, T. "Changes in serum and spinal fluid calcium effected by disodium ethylenediaminetetraacetate." *J. Lab. Clin. Med.*, 58:542–47. 1961. (10)

Soffer, A., Toribara, T., Moore-Jones, D., and Weber, D. "Clinical applications and untoward reactions of chelation in cardiac arrhythmias." *AMA Arch. Intern. Med.*, 106:824–34, 1960. (18)

Soffer, A., Toribara, T. and Sayman, A. "Myocardial responses to chelation." *Brit. Heart J.*, 23:690–94, 1961. (20)

Stacy, B. D., and Thorborn, G. D. "Chromium-51 ethylenediaminetetraacetate for estimation of glomerular filtration rate. *Science*, 152:1076–77, 1966. (14)

Stahlavska, A., and Malat, M. "Quantitative determination of EDTA and other polyamino acetic acids in urine and serum." *Clin. Chem. Acta.*, 41:181–86, 1972. (19)

Stamp, T. C. B. "[51]Cr-edetic-acid clearance and G. F. R." (Letters) *Lancet*, 2:1348, 1968. (4)

Stamp, T. C. B., Stacey, T. E., and Rose, G. A. "Comparison of glomerular filtration rate measurements using inulin, ^{51}Cr-EDTA, and a phosphate infusion technique." *Clin. Chim. Acta.*, 30:351–58, 1970. (20)

Stavem, P. and Berg, K. "A macromolecular serum component acting on platelets in the presence of EDTA-'platelet stain preventing factor.' " *Scand. J. Haemat.*, 10:202–208, 1973. (2)

Stecker, R. H. and Bennett, M. "Chelation in clinical otosclerosis." *AMA Arch. Otolaryng.*, 70:627–29, 1959. (8)

Stein, E. A., Hsiu, J. and Fischer, E. H. "Alpha-amylases as calcium-metalloenzymes. I. Preparation of calcium-free apoamylases by chelation and electro dialysis." *Biochemistry* 3:56–61, 1964. (18)

Steven, F. S. "The effect of chelating agents on collagen interfibrillar matrix interactions in connective tissue." *Biochem. Biophys. Acta.*, 140:522–28, 1967. (13)

Stewart, G. G., Kapsimalas, P., and Rappaport, H. "EDTA and urea peroxide for root canal preparation." *J. Am. Dent. Assoc.*, 78:335–38, 1969. (15)

Stinnett, J. D., Gilleland, Jr., H. E., and Eagon, R. G. "Proteins released from cell envelopes of *Pseudomonas aeruginosa* on exposure to ethylenediaminetetraacetate: comparison with dimethyl-formamide-extractable proteins." *J. Bact.*, 114:399–407, 1973. (35)

Sullivan, T. J. "Effect of manganese edetate on blood formation in rats." *Nature*, 186:87, 1960. (5)

Sullivan, T. J. "The effects of metallic edetates on the growth and blood formation of rats." *Arch. Int. Pharmacodyn.*, 124:225–36, 1960. (21)

Sundberg, M. W., Meares, C. F., Goodwin, D. A., and Diamanti, C. I. "Chelating agents for the binding of metal ions to macromolecules." *Nature*, 250:587–88, 1974. (9)

Sundberg, M. W., Meares, C. F., Goodwin, D. A., and Diamanti, C. I. "Selective binding of metal ions to macromolecules using bifunctional analogs of EDTA." *J. Med. Chem.*, 17:1304–1307, 1974. (15)

Surawicz, B. "Use of the chelating agent, EDTA, in digitalis intoxication and cardiac arrhythmias." *Progr. Cardiov. Dis.*, 2:432–43, 1960. (40)

Surawicz, B., MacDonald, M. G., Kaljob, V., Bettinger, J. C., Carpenter, A. A., Korson, L., and Starcheska, Y. K. "Treatment of cardiac arrhythmias with salts of ethylenediamine tetraacetic acid (EDTA)." *Am. Heart J.*, 58:493–503, 1959. (32)

Suso, F. A., and Edward, Jr., H. M. "Binding of EDTA, histidine and acetylsalicylic acid to zinc-protein complex in intestinal content, intestinal mucosa and blood plasma." *Nature*, 236:230–32, 1972. (10)

Suso, F. A., and Edward, Jr., H. M. "Ethylenediaminetetraacetic acid and ^{65}Zn binding by intestinal digesta, intestinal mucosa and blood plasma." *Proc. Soc. Exp. Biol. Med.*, 138:157–62, 1971. (9)

Swenerton, H., and Hurley, L. S. "Teratogenic effects of a chelating agent and their prevention by zinc." *Science*, 173:62–64, 1971. (16)

Szekely, P. and Wynne, N. A. "Effects of calcium chelation on digitalis-induced cardiac arrhythmias." *Brit. Heart J.*, 25:589–94, 1963. (26)

Tate, S. S. "Effect of metal ions and EDTA on the activity of rabbit liver fructose 1,6-diphosphatase." *Biochem. Biophys. Res. Commun.*, 24:662–67, 1966. (6)

Taylor, D. M., and Jones, J. D. "Effects of ethylenediaminetetraacetate and diethylene-triaminepentaacetate on DNA. Synthesis in kidney and intestinal mucosa of folate treated rates." *Biochem. Pharmacol.*, 21:3313–15, 1972. (21)

Teisinger, J. "Biochemical responses to provocative chelation by edetate disodium calcium." *Arch. Environ. Health*, 23:280–83, 1971. (11)

Temmerman, J. and Lebleu, B. "Evidence for the detachment of a ribonucleoprotein messenger complex from EDTA-treated rabbit reticulocyte polyribosomes." *Biochem. Biophys. Acta.*, 174:544–50, 1969. (11)

Thomas, Jr., L. J. "Ouabain contracture of frog heart: Ca^{45} movements and effect of EDTA." *Am. J. Physiol.*, 199:146–50, 1960. (16)

Thorell, J. I., and Lanner, A. "Influence of heparin-plasma, EDTA-plasma, and serum on the determination of insulin with three different radioimmunoassays." *Scand. J. Clin. Lab. Invest.*, 31:187–90, 1973. (15)

Tidball, C. S. and Lipman, R. I. "Enhancement of jejunal absorption of heparinoid by sodium ethylenediaminetetraacetate in the dog." *Proc. Soc. Exp. Biol. Med.*, 111:713-15, 1962. (10)

Toribara, T. Y. and Koval, L. "Determination of calcium in solutions containing ethylenediaminetetraacetic acid." *J. Lab. Clin. Med.*, 57:630-34, 1961. (6)

Trap-Jensen, J., Korsgaard, O., and Lassen, N. A. "Capillary permeability to human skeletal muscle measured by local injection of ^{51}Cr-EDTA and ^{133}Xe." *Scand. J. Clin. Lab. Invest.*, 25:93-99, 1970. (23)

Traub, Y. M., Samuel, R., Lubin, E., Lweitus, Z., and Rosenfeld, J. B. "A comparison between the clearances of inulan, endogenous creatinine and ^{51}Cr-EDTA." *ISR. J. Med. Sci.*, 9:487-89, 1973. (11)

Tritthart, H., MacLeod, D. P., Stierle, H. E. and Krause, H. "Effects of Ca-free and EDTA-containing Tyrode solution on transmembrane electrical activity and contraction in guinea pig papillary muscle." *Pflugers Arch.*, 338:361-76, 1973. (21)

Ts'ao, C.-H. "Platelet aggregation by ristocetin in EDTA plasma: extensive clumping with high concentrations of EDTA." *Haemostasis*, 1:315-19, 1972/73. (5)

Tunis, M. "The inhibitory action of EDTA on erythrocyte agglutination by lectins." *J. Immun.*, 95:876-79, 1965. (15)

Urey, J. C. and Horowitz, N. H. "Effects of EDTA on tyrosinase and L-amino-acid oxidase induction in *Neurospora crassa.*" *Biochem. Biophys. Acta.*, 132:300-309, 1967. (23)

Van Tol, A., Black, W. J. and Horecker, B. L. "Activation of rabbit muscle fruc tose diphosphatase by EDTA and the effect of divalent cations." *Arch. Biochem. Biophys.*, 151:591-96, 1972. (11)

Vanderdeelen, J. "Separation of metal-EDTA complexes by thin-layer chromatography." *J. Chromatogr.*, 32:521-22, 1969. (5)

Van Kley, H. and Claywell, C. S. "Evaluation of EDTA as the chelator in the biuret reagent." *Clin. Chem.*, 19:621-23, 1973.

Vedso, S. and Rud, C. "Determination of calcium in urine with EDTA by means of cation exchange resin." *Scand. J. Clin. Lab. Invest.*, 15:395-98, 1963. (6)

Virgilio, R. W., Homer, L. D., Herman, C. M., Moss, G. S., Lowery, B. D., and Schwartz, S. L. "Comparison of inulin and chromium-EDTA spaces in the nephrectomized baboon." *J. Surg. Res.*, 10:370-76, 1970. (16)

Vohra, P., Krantz, E. and Kratzer, F. H. "Formation constants of certain zinc-complexes by ion-exchange method." *Proc. Soc. Exp. Biol. Med.*, 121:422-25, 1966. (7)

Ward, W. W., and Fastiggi, R. J. "Binding of EDTA to DEAE-cellulose and its interference with protein determinations." *Anal. Biochem.*, 50:154-62, 1972. (15)

Waxman, H. S., and Brown, E. B. "Clinical usefulness of iron chelating agents." *Progr. Hematol.*, 6:338-73, 1969. (138)

Weber, K. M. "Changes in rat kidney glycogen content following injections of Na_2 (Ca-EDTA)." *Experientia*, 24:703-704, 1968. (14)

Weinberg, E. D. "Known and suspected role of metal coordination in actions of antimicrobial drugs." *Fed. Proc.*, 20 (Suppl. 10):132-36, September 1961. (38)

Weiner, M. "Factors influencing the clinical use of chelates in iron storage disease." *Ann. NY Acad. Sci.*, 119:789-96, 1964. (27)

Weiner, P. D., and Pearson, A. M. "Inhibition of rigor mortis by ethylenediamine tetraacetic acid." *Proc. Soc. Exp. Biol. Med.*, 123:185–87, 1966. (16)

Weiser, R. "Combinations of edetic acid and antibiotics in the treatment of rat burns infected with a resistant strain of *Pseudonomonas aeruginosa.*" *J. Infect. Dis.*, 128:566–69, 1973. (16)

Weiser, R., Asscher, A. W. and Wimpenny, J. "In vitro reversal of antibiotic resistance by ethylenediamine tetraacetic acid." *Nature*, 219:1365–66, 1968. (14)

Weiser, R., Wimpenny, J. and Asscher, A. W. "Synergistic effect of edetic-acid/antibiotic combinations on *Pseudomonas aeruginosa.*" *Lancet*, 2:619–20, 1969. (8)

Weiss, L. "Studies on cell deformability. III. Some effects of EDTA on sarcoma 37 cells." *J. Cell. Biol.*, 33:341–47, 1967. (23)

Welcher, F. J. *The analytical uses of ethylenediaminetetraacetic acid.* Princeton: D. Van Nostrand Co., Inc., 1958. (963)

West, J. J., Nagy, B., and Gergely, J. "The effect of EDTA on spectral properties of ATP-, ADP-, and ITP-G-actin." *Biochem. Biophys. Res. Commun.*, 29:611–16, 1967. (16)

Westerfield, W. W. "Effect of metal-binding agents on metalloproteins." *Fed. Proc.*, 20 (Suppl. 10): 158–78, September 1961. (298)

Whitaker, J. A., Austin, W., and Nelson, J. D. "Edathamil calcium disodium (Versenate) diagnostic test for lead poisoning." *Pediatrics*, 29:384–88, 1962. (13)

White, J. C. "Effects of ethylenediamine tetraacetic acid (EDTA) on platelet structure." *Scand. J. Haemat.*, 5:241–54, 1968. (23)

Wilder, L. W., De Jode, L. R., Milstein, S. W., and Howard, J. M. "Mobilization of atherosclerotic plaque calcium with EDTA utilizing the isolation-perfusion principle." *Surgery*, 52:793–95, 1962. (6)

Wilkinson, S. G. "The sensitivity of pseudomonads to ethylenediaminetetra-acetic acid." *J. Gen. Microbiol.*, 47:67–76, 1967. (42)

Wilkinson, S. G. "Studies on the cell walls of Pseudomonas species resistant to ethylenediaminetetraacetic acid." *J. Gen. Microbiol.*, 54:195–213, 1968. (71)

Williams, D. R. "Metals, ligands, and cancer." *Chem. Rev.*, 72:203–13, 1972. (108)

Williams, J. D., Matthews, G. A., and Judd, A. W. "Oral calcium disodium versenate in treatment of lead poisoning." *Brit. J. Indust. Med.*, 19:211–15, 1962. (20)

Wilson, L. A. "Chelation in experimental Pseudomonas keratitis." *Brit. J. Ophthal.*, 54:587–93, 1970. (12)

Winder, P. R., and Curtis, A. C. "Edathamil in the treatment of scleroderma and calcinosis cutis." *AMA Arch. Derm.*, 82:732–36, 1960. (15)

Wolf, H. U. "Effects of ethylenediamine-tetra-acetate and deoxycholate on kinetic constants of the calcium ion-dependent adenosine triphosphatase of human erythrocyte membranes." *Biochem. J.*, 130:311–14, 1972. (15)

Woods, S. M., Peters, H. A., and Johnson, S. A. M. "Chelation therapy in cutaneous porphyria. A review and report of a five-year recovery." *Arch. Derm.*, 84:920–27, 1961. (28)

Wooley, R. E., Schall, W. D., Eagon, R. G., and Scott, T. A. "Efficacy of EDTA-tris-lysozyme lavage in the treatment of experimentally induced *Pseudomonas aeruginosa* cystitis in the dog." *Am. J. Vet. Res.*, 35:27–29, 1974. (29)

Wynn, J. E., Riet, B. Van't, and Borzelleca, J. F. "The toxicity and pharmacodynamics of EGTA: oral administration to rats and comparisons with EDTA." *Toxicol. Appl. Pharmacol.*, 16:807–17, 1970. (6)

Yamashita, T., Soma, Y., Kovayashi, S., Sekine, T., Titani, K., and Narita, K. "The amino acid sequence at the active site of myosin A adenosine triphosphatase activated by EDTA." (Letters) *J. Biochem.* (Tokyo), 55:576–77, 1964. (6)

Yang, W. C. "The stimulatory effect of EDTA on cardiac mitochondrial respiration." *Biochem. Biophys. Res. Commun.*, 2:22–25, 1960. (4)

Ziegler, J. M., Batt, A. M., and Siest, G. "Effects of saccharo 1,4 lactone, EDTA and p-nitro-phenol on the UDP-glucuronyl-transferase activity of rat liver microsomes." *C. R. Soc. Biol.*, 167:685–88, 1973. (14)

Zinterhofer, L. J. M., Jatlow, P. I., and Fappiano, A. "Atomic absorption determination of lead in blood and urine in the presence of EDTA." *J. Lab. Clin. Med.*, 78:664–74, 1971. (10)

Zipkin, I., and Larson, R. H. "Caries potentiating effect of Na FDTA, Ca EDTA, and Mg EDTA in the rat." *J. Dent. Res.*, 38:1240, 1959. (3)

Zorzopulos, J., Jobbagy, A. J., and Terenzi, H. F. "Effects of ethylenediaminetetraacetate and chloramphenicol on mitochondrial activity and morphogenesis in *Mucor rouxii.*" *J. Bact.*, 115:1198–1204, 1973. (17)

Zydbek, G. J., and Kelly, Jr., E. W. "Preliminary investigation of the action of THQ, EDTA and estrogen on the mucopolysaccharides of sponge implant connective tissue." *J. Invest. Derm.*, 44:252–55, 1965. (10)

Index

About the Author

Morton Walker, D.P.M., is the author of 56 books on wholistic medicine, orthomolecular nutrition, and alternative methods of healing. He has worked full-time as a free-lance professional medical journalist and author since 1969. Prior to that, Dr. Walker was a practicing doctor of podiatric medicine for almost 17 years.

Dr. Walker has won 22 medical journalism awards, the most recent being the HUMANITARIAN AWARD presented to him by the Cancer Control Society in Pasadena, California, on September 6, 1992. The award cited Dr. Walker for his achievements as "The World's Leading Medical Journalist Specializing in Wholistic Medicine." The large bronze plaque, presented before an audience of 3,000, said, "You, Morton Walker, by Your Dedication, Competence, and Integrity Displayed Over More than a Quarter of a Century of Prolific Reporting, Continue to Earn the Respect and Gratitude of the Growing Thousands Who Are Benefiting from Alternative Approaches to Healing."

The Cancer Control Society in this first award it had ever made described well the journalistic/medical professionalism of Morton Walker, D.P.M.

Get a *SECOND OPINION* every month with Dr. Douglass' medical newsletter

Here's a shocker for you: Did you know that cancer, heart disease, the common cold, and a host of other "incurable" or "chronic" illnesses, are in many cases now completely reversible?

Did you know that garlic can help with certain forms of depression? That cabbage juice can cure the most stubborn painful ulcer — almost immediately? That extra magnesium in your diet can reduce tendencies toward anxiety, obesity, and even heart palpitations?

It's true. And it's exactly the kind of helpful medicine that can help keep you out of your doctor's office. It'll help you live longer, feel better, even look younger. You can only find such invaluable advice in SECOND OPINION.

With SECOND OPINION, you'll see your doctor a lot less ... and be much happier and healthier while you're at it. Go ahead and subscribe today! When you do, we'll give you one of the reports or books described on the next page absolutely free ... you choose the one you want!

Choose your free book/report on the next page.

Choose one of our special reports as your free gift!

AIDS
Why It's Much Worse Than They're Telling Us

AIDS: Why It's Much Worse Than They're Telling Us, And How To Protect Yourself And Your Loved Ones

Yes, AIDS is easy to catch. No, it isn't limited to just a few groups of society. People who've never engaged in questionable behavior or come within miles of an infected needle are contracting this deadly scourge. To protect yourself, you must know the truth.

Dangerous Drugs

If you knew what we know about the most popular prescription and over-the-counter drugs, you'd be sick. That's why Dr. Douglass wrote Dangerous Drugs. He gives you the low-down on 15 different categories of drugs: everything from painkillers and cold remedies to tranquilizers and powerful cancer drugs.

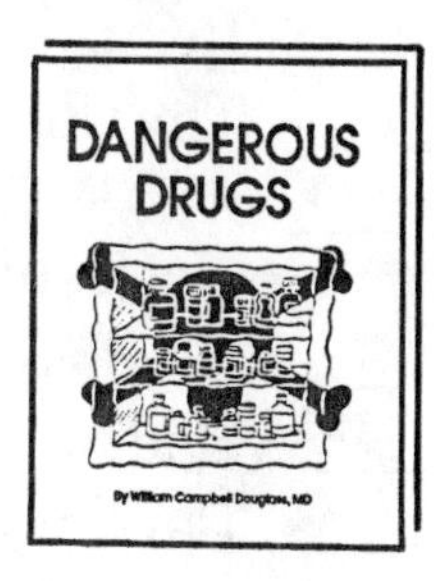

BAD MEDICINE

Bad Medicine

Do you really need that new prescription or that overnight stay in the hospital? In this report, Dr. Douglass reveals the common medical practices and misconceptions endangering your health. Best of all, he tells you the pointed (but very revealing!) questions your doctor prays you never ask!

Eat Your Cholesterol

Never feel guilty about what you eat again! Dr. Douglass shows you why red meat, eggs, and dairy products aren't the dietary demons we're told they are. But beware: This scientifically sound report goes against all the "common wisdom" about the foods you should eat. Read with an open mind!

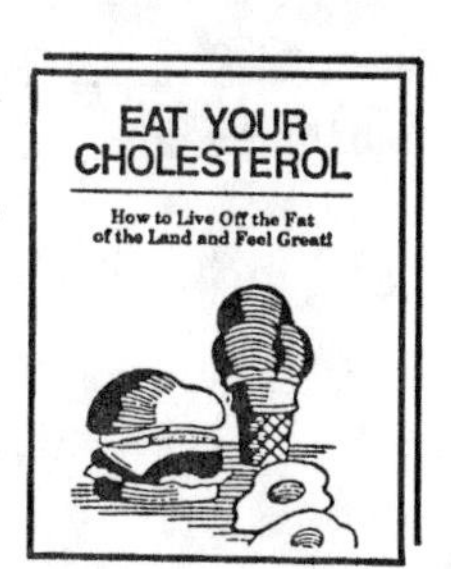

To subscribe and choose your free gift, please use the order form on the next page.

ORDER HERE

❑ **I'M SUBSCRIBING TO SECOND OPINION at just $49 for 12 issues (reg. $98 — I save 50%.)** I want to protect those I love from the health dangers the authorities aren't telling me about ... and the incredible cures that they've scorned and ignored. This kind of information is vital to my health. Sign me up for SECOND OPINION — and don't forget to send me my free gift. The free report/book I want is

____.

(choose 1 title from below)

I'd like to buy the following:

Qty.	Title	Price	Amount
____	Into the Light	$24.50	$______
____	The Chelation Answer	$16.95	$______
____	AIDS: What They're Not Telling You	$19	$______
____	Dangerous Drugs	$19	$______
____	Bad Medicine	$19	$______
____	Eat Your Cholesterol	$19	$______

Add shipping/handling per order:
$2.50 first item, .50 each additional item $______

TOTAL $______

❑ My payment of $__________ is enclosed.
❑ Charge my: ❑ MasterCard ❑ Visa

Card#______________________________________
Signature__________________________ Exp.__________

Name______________________________________
Address____________________________________
City________________________ State_____ Zip________
Telephone___________________________________

CHEL93

Call Toll-Free

1-800-728-2288

Fax: 404-399-0815

Mail to: **Second Opinion**
P.O. Box 467939 • Atlanta, GA 30346–7939